LAND USE INTENSIFICATION

Effects on Agriculture, Biodiversity and Ecological Processes

Editors:

David Lindenmayer
The Australian National University

Saul Cunningham
CSIRO Ecosystem Sciences

Andrew Young
CSIRO Plant Industry

CSIRO

PUBLISHING

CRC

National Library of Australia Cataloguing-in-Publication entry

> Land use intensification : effects on agriculture, biodiversity and ecological processes/edited by David Lindenmayer, Saul Cunningham and Andrew Young.

> 9780643104075 (pbk.)
> 9780643104082 (epdf)
> 9780643104099 (epub)

> Includes bibliographical references and index.

> Land use – Environmental aspects.
> Agricultural landscape management.
> Conservation of natural resources – Economic aspects.
> Ecosystem management – Economic aspects.

> Lindenmayer, David.
> Cunningham, Saul.
> Young, Andrew.

> 333.73137

Published exclusively in Australia and New Zealand by
CSIRO PUBLISHING
150 Oxford Street (PO Box 1139)
Collingwood VIC 3066
Australia

Telephone: +61 3 9662 7666
Local call: 1300 788 000 (Australia only)
Fax: +61 3 9662 7555
Email: publishing.sales@csiro.au
Web site: www.publish.csiro.au

Published in the rest of the world by
CRC Press
6000 Broken Sound Parkway NW Suite 300
Boca Raton, FL 33487 USA
Visit the CRC Press web site: http://www.crcpress.com

ISBN: 978-1-4665-1714-1

Front cover images by iStockphoto

Set in 10/12 Adobe Minion Pro and ITC Stone Sans
Edited by Anne Findlay, Editing Works Pty Ltd
Cover and text design by James Kelly
Typeset by Desktop Concepts Pty Ltd, Melbourne
Index by Russell Brooks
Printed in China by 1010 Printing International Ltd

CSIRO PUBLISHING publishes and distributes scientific, technical and health science books, magazines and journals from Australia to a worldwide audience and conducts these activities autonomously from the research activities of the Commonwealth Scientific and Industrial Research Organisation (CSIRO). The views expressed in this publication are those of the author(s) and do not necessarily represent those of, and should not be attributed to, the publisher or CSIRO.

Original print edition:
The paper this book is printed on is in accordance with the rules of the Forest Stewardship Council®. The FSC® promotes environmentally responsible, socially beneficial and economically viable management of the world's forests.

Contents

List of contributors

Simon J Attwood
Biodiversity Conservation Branch, Department of Sustainability, Environment, Water, Population and Communities, Canberra, ACT 2600, Australia

Péter Batáry
Agroecology, Department of Crop Sciences, Georg-August University, Grisebachstr. 6, 37077 Göttingen, Germany; MTA-ELTE-MTM Ecology Research Group Ludovika ter 2, 1083 Budapest, Hungary

Kamaljit S. Bawa
University of Massachusetts, Boston, MA 02125, USA; and Ashoka Trust for Research in Ecology and the Environment, Bangalore 660 024, India

Tim G. Benton
Institute of Integrative & Comparative Biology, University of Leeds, LS2 9JT

Linda Broadhurst
CSIRO Plant Industry, GPO Box 1600, Canberra, ACT 2601, Australia

Barry W. Brook
School of Earth and Environmental Sciences, University of Adelaide, SA 5005, Australia

Emma Burns
Biodiversity Conservation Branch, Department of Sustainability, Environment, Water, Population and Communities, Canberra, ACT 2600, Australia

Pashupati Chaudhary
University of Massachusetts, Boston, MA 02125, USA

Yann Clough
Agroecology, Department of Crop Sciences, Georg-August University, Grisebachstr. 6, 37077 Göttingen, Germany

Saul Cunningham
CSIRO Ecosystem Sciences, GPO Box 1600, Canberra, ACT 2601, Australia

Elizabeth L. Deakin
School of Biological Sciences, University of Canterbury, Private Bag 4800, Christchurch, New Zealand

Lisa H. Denmead
School of Animal Biology, The University of Western Australia, 35 Stirling Highway, Crawley, WA 6009, Australia

Raphael K. Didham
School of Animal Biology, The University of Western Australia, 35 Stirling Highway, Crawley, WA 6009, Australia; CSIRO Ecosystem Sciences, Centre for Environment and Life Sciences, Underwood Ave, Floreat, WA 6014, Australia; and School of Biological Sciences, University of Canterbury, Private Bag 4800, Christchurch, New Zealand

Don A. Driscoll
Fenner School of Environment and Society; ARC Centre of Excellence for Environmental Decisions; and National Environmental Research Program, The Australian National University, Canberra, ACT 0200, Australia

David H. Duncan
Arthur Rylah Institute for Environmental Research, Department of Sustainability and Environment, PO Box 137, Heidelberg, VIC 3084

Shristi Kamal
Ashoka Trust for Research in Ecology and the Environment, Bangalore 660 024, India

David Kleijn
Alterra, Centre for Ecosystem Studies, PO Box 47, 6700 AA, Wageningen, The Netherlands

Lian Pin Koh
Department of Environmental Sciences, ETH Zurich, CHN G 73.1 Universitatstrasse 16, 8092 Zurich, Switzerland

Janice S. H. Lee
Department of Environmental Sciences, ETH Zurich, CHN G 73.1 Universitatstrasse 16, 8092 Zurich, Switzerland

Tien Ming Lee
Ecology, Behaviour and Evolution Section, Division of Biological Sciences, University of California, San Diego, 9500 Gilman Drive, MC 0116, La Jolla, CA 92092, USA; and Department of Ecology and Evolutionary Biology, Yale University, New Haven, P.O. Box 208106, New Haven, CT 06520-8106, USA

David Lindenmayer
Fenner School of Environment and Society; ARC Centre of Excellence for Environmental Decisions; and National Environmental Research Program, The Australian National University, Canberra, ACT 0200, Australia

Sue McIntyre
CSIRO Ecosystem Sciences, PO Box 284, Canberra, ACT 2601, Australia

Kelvin S.-H. Peh
Department of Zoology, University of Cambridge, Downing Street, CB2 3EJ, UK

Ivette Perfecto
School of Natural Resources and Environment, University of Michigan, 440 Church St, Ann Arbor, MI 48109–1041, USA

Mary Rose C. Posa
Department of Biological Sciences, National University of Singapore, 14 Science Drive 4, Singapore 117543

Suman Rai
Ashoka Trust for Research in Ecology and the Environment, Bangalore 660 024, India

Michael J. Samways
Department of Conservation Ecology and Entomology, Stellenbosch University, P/Bag X1, Matieland 7602, South Africa

Christoph Scherber
Agroecology, Department of Crop Sciences, Georg-August University, Grisebachstr. 6, 37077 Göttingen, Germany

Navjot S. Sodhi
Department of Biological Sciences, National University of Singapore, 14 Science Drive 4, Singapore 117543

Malcolm C.K. Soh
NUS High School of Mathematics and Science, 20 Clementi Avenue 1, Singapore 129957

Carsten Thies
Agroecology, Department of Crop Sciences, Georg-August University, Grisebachstr. 6, 37077 Göttingen, Germany

Teja Tscharntke
Agroecology, Department of Crop Sciences, Georg-August University, Grisebachstr. 6, 37077 Göttingen, Germany

John Vandermeer
Department of Ecology and Evolutionary Biology, University of Michigan, 830 North University, Ann Arbor, MI 48109-1048, USA

Marc-André Villard

Département de Biologie, Université de Moncton, Moncton, NB E1A 3E9, Canada

Thomas C. Wanger

Center for Conservation Biology, Department of Biology, Stanford University, 371 Serra Mall, CA 94305-5020, USA; and Ecosystem Functions, Institute of Ecology, Leuphana University of Lüneburg, Scharnhorststraße 1, 21335 Lüneburg, Germany

Catrin Westphal

Agroecology, Department of Crop Sciences, Georg-August University, Grisebachstr. 6, 37077 Göttingen, Germany

Andrew Young

CSIRO Plant Industry, GPO Box 1600, Canberra, ACT 2601, Australia

1 LAND USE INTENSIFICATION: A CHALLENGE FOR HUMANITY

David Lindenmayer, Saul Cunningham and Andrew Young

Introduction

By 2050 the global population of humans is predicted to increase by 35% to ~9 billion and up to 70% more food will be required (Bruinsma 2009; Cribb 2010). Yet such demands for increased agricultural production will take place against a backdrop of already light to severe land degradation in 16–40% of existing agricultural areas (Chappell and Lavalle 2011), rapid reductions in available freshwater (Rockström et al. 2009), rapid climate change (Steffen et al. 2009) and its effects on global food harvest (Battisti and Naylor 2009), rapid expansion of new sources of energy like biofuels (Fargione et al. 2010), and major ongoing losses of biodiversity (Butchart et al. 2010), including in areas where the primary form of land use is agricultural production (Brown 2008; Millennium Ecosystem Assessment 2005).

This alignment of current and rapidly emerging problems represents a truly major set of environmental, economic and social challenges for humanity (Cribb 2010; Ehrlich and Ornstein 2010). Indeed, there is a massive literature already dedicated to examining this problem, which a quick web-based search in May 2011 suggests currently comprises over 3000 scientific articles! One widely debated approach to tackling problems like trade-offs between agricultural production and biodiversity conservation is land sparing in which more intensive land use is adopted in a given location as a kind of ecological offset for increasing the area reserved for conservation in another location (Fischer et al. 2008; Green et al. 2005; Phalan et al. 2011). However, there are other complicating dimensions to this approach such as: (1) the potential for the degradation of key ecosystem processes like pest control and pollination that underlie agricultural production (Garibaldi et al. 2011; Tscharntke et al. 2005), (2) the fact that some landscapes are already so degraded (Chappell and Lavalle 2011) there is little or no potential for land use intensification; and (3) there are some regions where there are few alternative landscapes to which the land sparing approach to biodiversity offsets can validly be applied (Gibbons and Lindenmayer 2007).

For all these reasons, land sparing is one of many controversial, yet important (and still unresolved) issues associated with complex multi-faceted land use intensification problems.

Given the magnitude of the problems facing humanity coupled with the enormous interest within the broader scientific community – as well as among the general public and politicians about these problems – we strongly believed that it was time to bring together people with different perspectives on the intersection of land use intensification with other values like biodiversity conservation. Land use intensification was therefore the central topic of a meeting held on the outskirts of Sydney in April 2011. This book's chapters are the product of that meeting and it comprises chapters from experts with markedly different perspectives from markedly different parts of the world. The contributions come from North America, Australia, Europe, United Kingdom, Asia, Africa and Latin America. They span the agricultural, natural forest and plantation sectors and cover different kinds of enterprises, from localised smallholdings to large, industrial-scale farms and forests.

Chapter and book structure

We decided at the early planning stages of the April 2011 meeting that the ideas and insights about land use intensification and its effects on biodiversity should be captured in a readily accessible book. This was because the valuable ideas and insights that have arisen in many past discussions in structured workshops and meetings have never been written up. We believe it is vital that insights about such a critically important topic as land use intensification are widely disseminated.

We provided each author with a template to guide their writing. This has given rise to the broadly consistent chapter structure of all of the contributions in this book. Specifically, we asked each author to address the question:

> 'What are the 5–6 key (i.e. most important) lessons that that have arisen from your work on biodiversity responses to land use intensification?'

Our initial concern with commissioning chapters for this book was that there would be enormous overlap in the contributions and we would be confronted with some 15 chapters of similar content, all calling for the same set of recommendations. This did not eventuate and the content of the chapters is as diverse as it is insightful and thought provoking. In fact, the diversity of the chapters made it difficult to determine an appropriate order of appearance throughout the book. After much thought, we elected to present the chapters as follows: the first set of chapters is broadly (but also loosely) grouped under 'General themes and principles' and based on insights and recommendations that are location-neutral. The second set of chapters is a collection of case studies closely tied to particular landscapes and/or industries. The final chapter in this book is a General Discussion. It is built around two components: (1) a general overview of the different perspectives that characterise the chapters in the preceding parts of the book – a standard part of most concluding chapters in any edited book; and (2) a synthesis of material derived from the extensive discussions at the meeting in April 2011.

Some caveats

We are acutely aware that only a small subset of those people with expertise in land use intensification was invited to attend our April 2011 meeting and contribute a chapter to this book. However, the meeting was necessarily small to allow sensible and tractable discussions (each presentation at the workshop could not exceed 15 minutes). Given this, we fully acknowledge that there will be other perspectives that have not been represented, either in part or in full, in this volume. In many respects this is good because it means there is a lot more to be said and written about land use intensification for improved conservation outcomes. If this book can stimulate additional dialogue and, in turn, foster more support for better-informed land use management, then we strongly believe that this exercise will have been a valuable one.

Overarching aims

This book has two aims. The first is to produce a short pithy book of case studies about relationships between biodiversity conservation and land use intensification. By capturing critical insights into successes, failures and solutions, we hope to provide high-level guidance for important initiatives around land use management. Ultimately, our second broad aim, which follows from the first one, is to seek ways to arrest the decline of biodiversity as a result of land use intensification.

David Lindenmayer, Saul Cunningham and Andrew Young
June 2011

References

Battisti DS and Naylor RL (2009) Historical warnings of future food insecurity with unprecedented seasonal heat. *Science* **323**, 240–244.

Brown LR (2008) *State of the World*. World Watch Institute and W.W. Norton, Washington, D.C.

Bruinsma J (2009) *The Resource Outlook to 2050: By How Much Do Land, Water and Crop Yields Need to Increase by 2050?* FAO, Rome, Italy.

Butchart SHM, Walpole M, Collen B, Van Strien A, Scharlemann JPW, Almond REA, Baillie JEM, Bomhard B, Brown C, Bruno J, Carpenter KE, Carr GM, Chanson J, Chenery AM, Csirke J, Davidson NC, Dentener F, Foster M, Galli A, Galloway JN, Genovesi P, Gregory RD, Hockings M, Kapos V, Lamarque J-F, Leverington F, Loh J, McGeoch MA, McRae L, Minasyan A, Morcillo MH, Oldfield TEE, Pauly D, Quader S, Revenga C, Sauer JR, Skolnik B, Spear D, Stanwell-Smith D, Stuart SN, Symes A, Tierney M, Tyrrell TD, Vié J-C and Watson R (2010) Global biodiversity: indicators of recent declines. *Science* **32**, 1164–1168.

Chappell MJ and Lavalle LA (2011) Food security and biodiversity: can we have both? *Agriculture and Human Values* **28**, 3–26.

Cribb J (2010) *The Coming Famine. The Global Food Crisis and What We Can Do to Avoid It*. CSIRO Publishing and University of California Press, Melbourne and Oakland.

Ehrlich PR and Ornstein RE (2010) *Humanity on a Tightrope*. Rowman and Littlefield Publishers, New York.

Fargione JE, Plevin RJ and Hill JD (2010) The ecological impact of biofuels. *Annual Review of Ecology, Evolution, and Systematics* **41**, 379–406.

Fischer J, Brosi B, Daily G, Ehrlich P, Goldman R, Goldstein J, Lindenmayer DB, Manning A, Mooney H, Pejchar L, Ranganathan J and Tallis H (2008) Should agricultural policies encourage land sparing or wildlife-friendly farming? *Frontiers in Ecology and the Environment* **6**, 380–385.

Garibaldi LA, Aizen MA, Klein AM, Cunningham SA and Harder LD (2011) Global growth and stability of agricultural yield decrease with pollinator dependence. *Proceedings of the National Academy of Sciences* **108**, 5909–5914.

Gibbons P and Lindenmayer DB (2007) Offsets for land clearing: no net loss or the tail wagging the dog? *Environmental Management & Restoration* **8**, 26–31.

Green RE, Connell SJ, Scharlemann JP and Balmford A (2005) Farming and the fate of wild nature. *Science* **307**, 550–555.

Millennium Ecosystem Assessment (2005) 'Millennium Ecosystem Assessment Synthesis Report'. At http://www.millenniumassessment.org.

Phalan B, Balmford A, Green RE and Scharlemann JPW (2011) Minimising harm to biodiversity of producing more food globally. *Food Policy* **36**, S62–S71.

Rockström J, Steffen W, Noone K, Persson Å, Chapin IFS, Lambin E, Lenton TM, Scheffer M, Folke C, Schellnhuber H, Nykvist B, De Wit CA, Hughes T, van der Leeuw S, Rodhe H, Sörlin S, Snyder PK, Costanza R, Svedin U, Falkenmark M, Karlberg L, Corell RW, Fabry VJ, Hansen J, Walker B, Liverman D, Richardson K, Crutzen P and Foley J (2009) Planetary boundaries: exploring the safe operating space for humanity. *Ecology and Society* **14**, 32.

Steffen W, Burbidge A, Hughes L, Kitching R, Lindenmayer DB, Musgrave W, Stafford-Smith M and Werner P (2009) *Australia's Biodiversity and Climate Change*. CSIRO Publishing, Melbourne.

Tscharntke T, Klein AM, Kruess A, Steffan-Dewenter I and Thies C (2005) Landscape perspectives on agricultural intensification and biodiversity-ecosystem service management. *Ecology Letters* **8**, 857–874.

Part A

General themes and principles

2 COMBINING BIODIVERSITY CONSERVATION WITH AGRICULTURAL INTENSIFICATION

Teja Tscharntke, Péter Batáry, Yann Clough, David Kleijn, Christoph Scherber, Carsten Thies, Thomas C. Wanger and Catrin Westphal

Lesson #1: Pesticides are a largely underestimated determinant of biodiversity loss.

Lesson #2: Farmland biodiversity reduces household vulnerability and provides natural insurance to risk-averse farmers.

Lesson #3: Biodiversity conservation needs a landscape perspective.

Lesson #4: Farmland biodiversity is good for ecosystem services but rarely includes endangered species.

Lesson #5: High yield and high farmland biodiversity can be combined.

Lesson #6: The concept of land sparing, instead of wildlife-friendly farming, does not contribute to connecting hunger reduction with biodiversity conservation.

Introduction

Conversion of natural habitat and agricultural intensification are the most important drivers of global losses in biodiversity and associated processes. Paradoxically, agricultural intensification at local and landscape scales tends to make land use systems less resilient and more vulnerable to disturbances while environmental change and climate extremes call for a higher adaptation capacity than ever. Biodiversity loss means that ecosystem services are also endangered, affecting functioning of managed and natural ecosystems. Conservation of the biodiversity inside protected areas receives increasing attention, but management of human-dominated landscapes, including forest remnants and forested land use systems, is still a major challenge. In this chapter, we focus primarily on the relationship between biodiversity and agricultural land use.

Lessons

1. Pesticides are a largely underestimated determinant of biodiversity loss

Agricultural intensification has many components, such as loss of landscape elements, enlarged farm and field sizes and larger local inputs of fertiliser and pesticides (Tscharntke *et al.* 2005). High pesticide use can enhance crop production, but may also damage human health, reduce biodiversity and impair ecosystem services. In a Europe-wide study, the use of insecticides, herbicides and fungicides proved to be the best predictor of losses in plant, bird and carabid beetle richness as well as reduced biological control (Geiger *et al.* 2010). The non-target effects of pesticides on tropical biodiversity and people are largely unknown, with less than 0.1% of all pertinent pesticide papers dealing with tropical vertebrate diversity. However, amphibians (with their sensitive skin), for example, experience higher pesticide susceptibility in tropical than in temperate regions. Specifically, pesticide-mediated indirect effects on biodiversity and ecosystem services such as pollination and biological control require urgent research attention in the near future.

2. Farmland biodiversity reduces household vulnerability and provides natural insurance to risk-averse farmers

Diverse agroecosystems are characterised by a high natural insurance function against changing environments because they decrease the variance of crop yields and, thereby, the uncertainty in the provision of public-good ecosystem services (Baumgärtner and Quaas 2010). For instance, imagine agroecosystems without the hundreds of predatory species that control many plant-feeding arthropods that could otherwise become major pests (Tscharntke *et al.* 2007). And then imagine severe losses in the numbers of pollinators, which are known to influence reproduction of most wild plant species and a third of the global food production (Klein *et al.* 2007). For example, yields of highland coffee can increase by 50% when pollinated by bees; cacao is completely dependent on cross-pollination; and yet the biology of pollinating midges is little understood (Groeneveld *et al.* 2010). Added to this, the genetic diversity of crops is important for pest and disease management, pollination services and soil processes (Hajjar *et al.* 2008).

Adaptation strategies to environmental change include the maintenance of shade trees in tropical agroforestry, but conversion from shaded to unshaded systems is common practice to increase short-term yield (Perfecto *et al.* 2007). Shade trees in agroforestry enhance functional biodiversity, carbon sequestration, soil fertility and drought resistance as well as providing biological weed and pest control. Shade is important for young trees, but is less so for older cacao trees, so farmers often remove shade trees in older cacao plantations (Tscharntke *et al.* 2011). In a 'long-term cacao boom-and-bust cycle', a cacao boom can be followed by a cacao bust due to unmanageable pest and pathogen levels (such as in Brazil and Malaysia). Presumably, this is the result of physiological stress suffered by unshaded cacao as well as the pests accumulating in the larger areas used for cacao production. Risk-averse farmers (for example in Bahia, Brazil) avoid long-term vulnerability of their agroforestry systems by keeping shade as an insurance against insect pest outbreaks, whereas yield-maximising farmers (for example in Indonesia) reduce shade and aim to achieve short-term monetary benefits (Tscharntke *et al.* 2011; see Figure 2.1). A better understanding of how organisms and their interactions contribute to essential ecosystem services would greatly reduce the need for external agrochemical inputs and increase both agricultural productivity and sustainability in temperate and tropical systems.

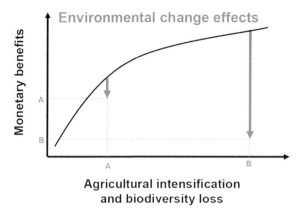

Figure 2.1: Conceptual model illustrating increasing vulnerability to environmental change with agricultural intensification. For example, shade tree loss in cacao agroforestry as an indicator of agricultural intensification. Monetary benefits (households' income stability) decrease less in low intensity (A) than high intensity production systems (B) (from Tscharntke *et al.* 2011).

3. Biodiversity conservation needs a landscape perspective

In agricultural landscapes, species experience their environment across a range of spatial scales. Landscape composition and configuration affect colonisation–extinction dynamics and biodiversity patterns. Key drivers of agroecosystem processes often come from the outside, that is the surrounding landscape matrix. The spillover of species between managed and natural ecosystems is an important process affecting cropland and wildlife populations alike (Bianchi *et al.* 2006; Rand *et al.* 2006). The spatio-temporal stability of resource availability makes natural habitats a well-recognised source of populations whose individuals move into managed systems. Conversely, the often high productivity of crop fields, as well as temporal pulses in their resource availability from growing crops until harvest, can at times make them a source of organisms spilling over to adjacent wild habitat. While the latter phenomenon is likely to occur frequently in production land-scapes, it has been surprisingly little studied to date (Rand *et al.* 2006; Holzschuh *et al.* 2011).

The intermediate landscape-complexity hypothesis states that the effectiveness of local conservation management, such as organic farming or implementing boundary strips, is highest in structurally simple, rather than in cleared (i.e. extremely simplified) or in complex landscapes (Batáry *et al.* 2011).

4. Farmland biodiversity is good for ecosystem services but rarely includes endangered species

A matrix of wildlife-friendly agroecosystems and natural habitat patches enhances disper-sal, and therefore the survival of populations (Tscharntke *et al.* 2005; Perfecto *et al.* 2009). Many small habitat remnants across a large geographic area protect more species than a single large remnant of the same area (Tscharntke *et al.* 2002). Nevertheless, fragmented populations experience high extinction risks, and many of the most endangered plants and animals need very large areas to survive.

A mosaic of managed and natural habitat can maximise pollination and yield (Klein *et al.* 2007; Priess *et al.* 2007) as well as biological pest control (Thies and Tscharntke 1999; Tscharntke *et al.* 2007), but such multifunctional landscapes only allow long-term survival of those species that are adapted to these land use systems. For example, bird diversity can be high in modified tropical landscapes, but endangered forest species are rare (Daily *et al.* 2001; Maas *et al.* 2009). Similarly, reducing shade in cacao agroforestry from 80% to 40% still supports both high yield and high biodiversity, but most forest species are lost (Steffan-Dewenter *et al.* 2007).

Conservation initiatives with intrinsic biodiversity objectives should therefore focus on remnants of natural habitat and, with respect to agriculture, to extensively used systems and structurally complex regions (Kleijn *et al.* 2011). In contrast, conservation of functionally important biodiversity providing major services such as pollination and biological control should target more on intensively farmed areas, because of higher potential benefits such as improved crop yields, reduced household vulnerability and fewer negative externalities from agrochemical use (Kleijn *et al.* 2011).

5. High yield and high farmland biodiversity can be combined

Wildlife-friendly farming has been criticised for being ineffective and rendering only low yields. Hence, the biodiversity–yield relationship is a critical issue in the discussion of whether wildlife-friendly farming is really desirable. Phalan *et al.* (2011) argue that yields are generally lower under wildlife-friendly farming than conventional management. However, at the global scale the reverse may be the case because in tropical regions biodiverse ecosystems are generally more productive (Badgley *et al.* 2007). Halweil (2006) argues that the eventual 20% increase in conventional compared to organic food production can be found only in high-tech agriculture with major agrochemical applications typically employed in wealthy nations. Such land use management is not available to many poor and hungry people, who rely on robust, secure and sustainable practices applied under an agro-ecological farming style (de Schutter 2011).

Biodiversity–yield relationships greatly differ depending on the regional context. In Europe, crop yield and biodiversity are usually negatively related (Kleijn *et al.* 2009; Geiger *et al.* 2010). Because the relationship is non-linear, reduction in land use intensity means fewer costs and higher impact on biodiversity in low compared to high intensity agroecosystems (Kleijn *et al.* 2011). In contrast, the yield–biodiversity relationship can be less clear in the tropics, where both biodiversity conservation and hunger mitigation is a major concern. In Indonesia, biodiversity and cacao crop productivity of smallholders were not related, opening substantial opportunities for wildlife-friendly management (Clough *et al.* 2011). Species richness of trees, fungi, invertebrates and vertebrates did not decrease with yield. Moderate shade, adequate labour and agricultural inputs can be combined with a complex habitat structure to enhance biodiversity as well as high yields. Therefore, at least in some situations, agroecosystems such as cacao agroforests can be designed to optimise both biodiversity and crop production benefits without adding pressure to convert natural habitat to farmland (Clough *et al.* 2011).

Current estimates of population and socio-economic growth foresee a doubling of food demand by 2050 (FAO 2008). Furthermore, the industry is increasingly looking for renewable plant-based resources to replace fossil resources. The rise of the bio-based economy will result in an increased demand for agricultural products. The challenge will be to find management regimes and landscape configurations that optimise functional biodiversity while maintaining agricultural efficiency and high productivity.

6. The concept of land sparing, instead of wildlife-friendly farming, does not contribute to connecting hunger reduction with biodiversity conservation

Combining efficient agricultural land use with biodiversity conservation is a challenge. Phalan *et al.* (2011) pose the question whether farming and conservation policies should be separated, segregating land for nature from land for production (land sparing), or integrated with production and conservation on the same land (wildlife-friendly farming). Because a high percentage of wild species cannot survive in even the most wildlife-friendly farming systems, protection of wild land is essential (Phalan *et al.* 2011). These authors conclude that agricultural intensification, thereby globally restricting human requirements for land, will be important in limiting the impacts of increasing food production on biodiversity.

So far so good, but such conclusions on maximising biodiversity and yield do not take into account the complexity of the real world. In developed countries, increasing yields are not related to decreasing areas in cropland (Ewers *et al.* 2009). In developing countries, higher agricultural and timber yields in a region attract migrants (at least in the long run) and enhance tropical deforestation, as shown by seminal reviews of socio-economists (Angelsen and Kaimowitz 1999). In addition, the long-term stability of food production and agroecosystem resilience are often weakened by intensified agricultural practices (de Schutter 2011; Tscharntke *et al.* 2011 and see above). Agricultural landscapes also have non-monetary values, including cultural goods and services increasing human wellbeing such as recreation.

Current global food production is sufficient to feed the world's seven billion people, which is in contrast to the fact that more than one billion people are malnourished today with poverty being the major cause of malnutrition (Chappell and LaValle 2011). There is enough food, even under scenarios of human population growth up to nine billion people, but the poor do not have the money to buy it (Chappell and LaValle 2011).

Quantification of food availability also has to take into account the inefficiency of our current ways of food production and consumption (e.g. Chappell and LaValle 2011; de Schutter 2011): (1) about one-third of global cereal production is used as feed for cattle, with eight calories of grain producing just one calorie of cattle meat; (2) post-harvest losses from pests account for about a third of the yield; and (3) about 30–50% of all food is thrown away uneaten. Woitowitz (2007) quantified the future food requirements for Germany based on the diet recommendations by German health institutions to reduce meat consumption to a third of its present value (from 60 kg to 20 kg per person and year). This would mean a 60% reduction of area for animal husbandry, the halving of greenhouse gas emissions (Woitowitz 2007), and providing more than enough space for wildlife-friendly farming.

Conclusions

Biodiversity within agroecosystems is essential for the maintenance of ecosystem services, whereas many endangered species often rely on large areas of natural habitat and cannot be protected within agricultural land use systems alone. However, half of our terrestrial environment is formed by agriculture and balancing human and ecological needs is more than ever a pressing issue in a densely populated world. Strategies of feeding a growing human population are not necessarily a trade-off between crop yield and biodiversity. Wildlife-friendly, organic farming can be very efficient, in particular in the tropics, where most of the poor and hungry people live and where win-win solutions (or synergies) appear to be a possible solution. In addition, economically viable agricultural practices in both temperate and tropical regions exhibit high variability, showing that there are different options available for promoting biodiversity. Combining agriculture with biodiversity conservation is a complex topic, a great challenge for the future and far from the simple solutions that the current land sparing versus wildlife-friendly farming debate suggests.

Acknowledgement

Financial support came from the German Ministry of Research and Education (BMBF) and the German Research Foundation (DFG).

Biographies

Teja Tscharntke holds a professorship in agroecology with a focus on the following topics: biodiversity patterns and associated ecosystem services such as pollination and biological control on different spatial and temporal scales and in tropical and temperate communities, quantified food webs and multitrophic interactions, and multidisciplinary approaches linking socioeconomic with ecological approaches. He is editor-in-chief of the journal *Basic and Applied Ecology*.

Péter Batáry holds a PhD in Biological Sciences and has joined the Agroecology Group in Göttingen with an Alexander-von-Humboldt fellowship. He was a Bolyai Research Fellow of the Hungarian Academy of Sciences. His research focus is on the effects of agri-environmental management of grasslands and arable fields on biodiversity, population ecology and ecological functioning, including analyses of landscape composition and configuration.

Yann Clough's research aims at increasing our understanding of how plant and animal communities are shaped in tropical and temperate landscapes dominated by humans, and how this feeds back into the support of human livelihoods through ecosystem services and disservices.

David Kleijn is a senior scientist with a research focus on biodiversity conservation on farmland. He explores causes of biodiversity decline, evaluates the impact of conservation management and policies and devises sustainable conservation strategies.

Christoph Scherber is a postdoctoral scientist interested in changes in biodiversity on ecosystem processes such as plant–insect interactions and herbivory. In addition, he is studying anthropogenic environmental changes (climate and habitat changes) in a variety of ecosystems such as heathland, peat bog ecosystems and agricultural landscapes.

Carsten Thies is a senior scientist and his current research activities focus on biodiversity and ecosystem functioning in agricultural landscapes. He has a strong background in agriculture, forestry and natural resource management, and works on the evaluation of environmental schemes and their implementation into current agri-environmental policies.

Thomas C. Wanger is a postdoctoral research fellow interested in tropical ecology. He is currently working on the effects of pesticide use on biodiversity, ecosystem services and human health. He also uses databases to understand which ecosystem services mammals and amphibians can provide on a global scale.

Catrin Westphal is a postdoctoral scientist with a focus on environmental drivers, such as climate change and land use intensification, and their effects on biodiversity and ecosystem functioning, in particular pollinators and pollination services. She is also interested in landscape and spatial ecology with a focus on landscape composition and configuration.

References

Angelsen A and Kaimowitz D (1999) Rethinking the causes of deforestation: lessons from economic models. *The World Bank Observer* **14**, 73–98.

Badgley C, Moghtader J, Quintero E, Zakem E, Chappell MJ, Avilés-Vázquez K, Samulon A and Perfecto I (2007) Organic agriculture and the global food supply. *Renewable Agricultural and Food Systems* **22**, 86–108.

Batáry P, Báldi A, Kleijn D and Tscharntke T (2011) Landscape-moderated biodiversity effects of agri-environmental management: a meta-analysis. *Proceedings of the Royal Society B* **278**, 1894–1902.

Baumgärtner S and Quaas MF (2010) Managing increasing environmental risks through agrobiodiversity and agrienvironmental policies. *Agricultural Economics* **41**, 483–496.

Bianchi FJJA, Booij CJH and Tscharntke T (2006) Sustainable pest regulation in agricultural landscapes: a review on landscape composition, biodiversity and natural pest control. *Proceedings of the Royal Society B* **273**, 1715–1727.

Chappell MJ and LaValle LA (2011) Food security and biodiversity: can we have both? *Agriculture and Human Values* **28**, 3–26.

Clough Y, Barkmann J, Juhrbandt J, Kessler M, Wanger TC, Anshary A, Buchori D, Cicuzza D, Darras D, Dwi Putra D, Erasmi S, Pitopang R, Schmidt C, Schulze CH, Seidel D, Steffen-Dewenter I, Stenchly K, Vidal S, Weist M, Wielgoss AC and Tscharntke T (2011) Combining high biodiversity with high yields in tropical agroforests. *Proceedings of National Academy of Sciences* **108**, 8311–8316.

Daily GC, Ehrlich PR and Sanchez-Azofeifa GA (2001) Countryside biogeography: use of human-dominated habitats by the avifauna of southern Costa Rica. *Ecological Applications* **11**, 1–13.

De Schutter O (2011) 'Report on the right to food'. UN Human Rights Council. http://www.srfood.org/images/stories/pdf/officialreports/20110308_a-hrc-16-49_agroecology_en.pdf

Ewers RM, Scharlemann JPW, Balmford A and Green RE (2009) Do increases in agricultural yield spare land for nature? *Global Change Biology* **15**, 1716–1726.

FAO (2008) 'Climate change adaptation and mitigation in the food and agriculture sector.' Technical background document from the expert consultation held on 5 to 7 March 2008, presented at Conference of Climate Change, Energy and Food. 3–5 June Rome, 2008. HLC/08/BAK1. ftp://ftp.fao.org/docrep/fao/meeting/013/ai782e.pdf.

Geiger F, de Snoo GR, Berendse F, Guerrero I, Morales MB, Onate JJ, Eggers S, Pårt T, Bommarco R, Bengtsson L, Clement LW, Weisser WW, Olszewski A, Ceryngier P, Hawro V, Inchausti P, Fischer C, Flohre A, Thies C and Tscharntke T (2010) Persistent negative effects of pesticides on biodiversity and biological control potential on European farmland. *Basic and Applied Ecology* **11**, 97–105.

Groeneveld JH, Tscharntke T, Moser G and Clough Y (2010) Experimental evidence for stronger cacao yield limitation by pollination than by plant resources. *Perspectives in Plant Ecology, Evolution and Systematics* **12**, 183–191.

Hajjar R, Jarvis DE and Gemmill-Herren B (2008) The utility of crop genetic diversity in maintaining ecosystem services. *Agriculture, Ecosystems and Environment* **123**, 261–270.

Halweil B (2006) Can organic farming feed the world? *World Watch* **19**, 18–24.

Holzschuh A, Dormann CF, Tscharntke T and Steffan-Dewenter I (2011) Expansion of mass-flowering crops leads to transient pollinator dilution and reduced wild plant pollination. *Proceedings of the Royal Society B Biological Sciences* **278**, 3444–3451.

Kleijn D, Kohler F, Báldi A, Batáry P, Concepción ED, Clough Y, Díaz M, Gabriel D, Holzschuh A, Knop E, Kovács A, Marshall EJP, Tscharntke T and Verhulst J (2009) On the relationship between farmland biodiversity and land-use intensity in Europe. *Proceedings of the Royal Society B* **276**, 903–909.

Kleijn D, Rundlöf M, Scheper J, Smith HG and Tscharntke T (2011) Does conservation on farmland contribute to halt biodiversity decline? *Trends in Ecology and Evolution* **26**, 474–481.

Klein A-M, Vaissière BE, Cane JH, Steffan-Dewenter I, Cunningham SA, Kremen C and Tscharntke T (2007) Importance of pollinators in changing landscapes for world crops. *Proceedings of Royal Society B* **274**, 303–313.

Maas B, Dwi Putra D, Waltert M, Clough Y, Tscharntke T and Schulze CH (2009) Six years of habitat modification in a tropical rainforest margin of Indonesia do not affect bird diversity but endemic forest species. *Biological Conservation* **142**, 2665–2671.

Perfecto I, Armbrecht I, Philpott SM, Soto-Pinto L and Dietsch TM (2007) Shaded coffee and the stability of rainforest margins in northern Latin America. In: *The Stability of Tropical Rainforest Margins, Linking Ecological, Economic and Social Constraints of*

Land Use and Conservation. (Eds T Tscharntke, C Leuschner, M Zeller, E Guhadja and A Bidin), pp. 227–264. Environmental Science Series, Springer Verlag, Berlin.

Perfecto I, Vandermeer J and Wright A (2009) *Nature's Matrix: Linking Agriculture, Conservation and Food Sovereignty.* Earthscan, London.

Phalan B, Balmford A, Green RE and Scharlemann JPW (2011) Minimising harm to biodiversity of producing more food globally. *Food Policy* **36**, S62–S71.

Priess JA, Mimler M, Klein AM, Schwarze S, Tscharntke T and Steffan-Dewenter I (2007) Linking deforestation scenarios to pollination services and economic returns in coffee agroforestry systems. *Ecological Applications* **17**, 407–417.

Rand TA, Tylianakis JM and Tscharntke T. (2006) Spillover edge effects: the dispersal of agriculturally subsidized insect natural enemies into adjacent natural habitats. *Ecology Letters* **9**, 603–614.

Steffan-Dewenter I, Kessler M, Barkmann J, Bos MM, Buchori D, Erasmi S, Faust H, Gerold G, Glenk K, Gradstein SR, Guhardja E, Harteveld M, Hertel D, Höhn P, Kappas M, Köhler S, Leuschner C, Maertens M, Marggraf R, Migge-Kleian S, Mogea J, Pitopang R, Schaefer M, Schwarze S, Sporn SG, Steingrebe A, Tjitrosoedirdjo SS, Tjitrosoemito S, Twele A, Weber R, Woltmann L, Zeller M and Tscharntke T (2007) Tradeoffs between income, biodiversity, and ecosystem functioning during tropical rainforest conversion and agroforestry intensification. *Proceedings of National Academy of Sciences* **104**, 4973–4978.

Thies C and Tscharntke T (1999) Landscape structure and biological control in agroecosystems. *Science* **285**, 893–895.

Tscharntke T, Steffan-Dewenter I, Kruess A and Thies C (2002) Contribution of small habitat fragments to conservation of insect communities of grassland-cropland landscapes. *Ecological Applications* **12**, 354–363.

Tscharntke T, Klein AM, Kruess A, Steffan-Dewenter I and Thies C (2005) Landscape perspectives on agricultural intensification and biodiversity-ecosystem service management. *Ecology Letters* **8**, 857–874.

Tscharntke T, Bommarco R, Clough Y, Crist TO, Kleijn D, Rand TA, Tylianakis JM, van Nouhuys S and Vidal S (2007) Conservation biological control and enemy diversity on a landscape scale. *Biological Control* **43**, 294–309.

Tscharntke T, Clough Y, Bhagwat SA, Buchori D, Faust H, Hertel D, Hölscher D, Juhrbandt J, Kessler M, Perfecto I, Scherber C, Schroth G, Veldkamp E and Wanger TC (2011) Multifunctional shade-tree management in tropical agroforestry landscapes – a review. *Journal of Applied Ecology* **48**, 619–629.

Woitowitz A (2007) Auswirkungen einer Einschränkung des Verzehrs von Lebensmitteln tierischer Herkunft auf ausgewählte Nachhaltigkeitsindikatoren – dargestellt am Beispiel konventioneller und ökologischer Wirtschaftsweise. Dissertation am Lehrstuhl für Wirtschaftslehre des Landbaus der Technischen Universität München.

3 MANAGING BIODIVERSITY IN AGRICULTURAL LANDSCAPES: PERSPECTIVES FROM A RESEARCH–POLICY INTERFACE

Simon J. Attwood and Emma Burns

Lesson #1. Land use change has enormous implications for biodiversity – both positive and negative.

Lesson #2. The history of agricultural development within a region can have a significant effect on biodiversity.

Lesson #3. Land use history greatly influences regional or national conservation priorities and approaches – in Australia and elsewhere. There is a need to better combine conservation in native vegetation and conservation in the agricultural matrix.

Lesson #4. Conservation initiatives should be informed by good science, but there are systemic obstacles to policy–research integration.

Lesson #5. Scientifically underpinned conservation incentive programs provide an opportunity for better integration and collaboration between researchers, policy makers and program officers. This can lead to a better knowledge base for all and improved conservation outcomes.

Introduction

Intensified land use change in agricultural landscapes can occur in many guises. This includes the replacement of native vegetation by crops or introduced pasture, the intensified management of land already cleared for agriculture, and the modification of native vegetation through activities such as livestock grazing. In general, such transitions have a deleterious impact upon biodiversity (Sala *et al.* 2000), although responses can vary considerably depending upon biological entities such as taxonomic group, life history traits, morphology, and geo-political features such as land use history (Attwood *et al.* 2008; Attwood 2010). There are marked differences in the ways that species and communities use different elements of agricultural landscapes (from intact, unmodified native vegetation to intensively managed production land) in different parts of the world. However, there is an increasing recognition and understanding of the need to adopt an approach to conservation that is more landscape-oriented, one that seeks to operate on all aspects of the landscape, from remnants of native vegetation to elements of the agricultural matrix. This shift in thinking has been driven by a vast body of scientific support and associated policy developments.

A common mechanism for delivering policy concerned with biodiversity decline in agricultural landscapes is to provide incentives to land managers to undertake certain conservation actions on their properties. There is a clearly identified need for such policy initiatives to be informed by good science. These initiatives may also present opportunities to guide research activities and priorities. There is willingness in both research and policy arenas to engage in increased collaboration and mutual information exchange (e.g. Commonwealth of Australia 2008). However, various obstacles hinder interaction. The use of incentive schemes to either prevent or manage the impacts upon biodiversity of land use change is one mechanism that offers important opportunities for greater interaction between research, policy development, program design and implementation.

Lessons

1. Land use change has enormous implications for biodiversity – both positive and negative

Land use change is arguably the greatest threat to biodiversity in the 21st Century (Sala *et al.* 2000). It is becoming increasingly determined by the global interconnectedness of economic and trade systems (Lambin *et al.* 2003), and hence may be less easily regulated at the regional and national scale. Furthermore, change is likely to accelerate in response to projected climate change, global human population increases, increasing food demands, shifting food consumption patterns in developing nations, and growing emphasis on food security (de Schutter 2010).

Ecologists and conservationists tend to think of land use change in terms of intensification and consequent biodiversity losses; it is certainly a sound rule of thumb in ecology that if one intensifies the management and/or simplifies the composition and structure of an agricultural landscape, then this often leads to large-scale loss of biodiversity (Fischer *et al.* 2008). However, losses are not equal between land use transition or between taxonomic or 'functional' groups. This is not a trivial matter – such discrepancies in biodiversity loss among land use types drive much of the ongoing debate between 'land sparing' and wildlife-friendly farming', two apparently competing paradigms of optimal biodiversity conservation in agricultural landscapes. To illustrate this, Attwood *et al.* (2008) found that losses of arthropod richness were greater between native vegetation and agricultural systems (e.g. woodland to introduced pasture) than between agricultural land use types (e.g. introduced pasture and cropping). Unpublished data by the same authors also found that while ant species richness declined from native woodland to semi-improved pastures in southern Queensland, several species of ant were present only in pastures, indicating that this land use may have considerable biodiversity value.

There may also be positive biodiversity outcomes from land abandonment (e.g. old fields) and ecological restoration sites (Cunningham *et al.* 2008; Lindenmayer *et al.* 2010).

Agri-environmental timeline for British Isles 10,000 yrs B.P. to present

10,000 - 5,500 BP	5,000 BP - 500 AD	500 - 1700 AD	1700 - 1900 AD	1900 - 1950 AD	1950 AD - present
Hunter gatherers Initial vegetation clearance	Agriculture dominant way of life Metal technology Well planned fields and wood pastures Clearing of uplands By Iron Age, 50% of woodlands cleared Large mammals declined	Well ordered rural landscape Loss of woodlands, heathlands, wetlands Wildwood replaced by managed woodland New crops, legumes and fertilisers increase yields Large mammals extinct	Fossil fuel dominance 6 million ha of arable land by 1870 Steam ploughs Enclosure Act Loss of wetlands and heathland Hedgerows increase	World wars lead to increased mechanisation Grazing intensified Agro-chemical use Government price intervention	Agricultural intensification, > field size, hedge removal, industrialised agriculture Production subsidies Homogenised countryside Biodiveristy loss (e.g. farmland birds) Agri-environment schemes Rise of organic movement

Agri-environmental timeline for Australia 60,000 yrs B.P. to present

60,000 - 10,000 BP	10,000 BP	4,000 BP - 18th Century	Late 1800s - 1900 AD	1900 - 1950 AD	1950 AD - present
First humans Climate change Shift from coastal to inland hunter gatherers Natural responses of vegetation to climate change Extinction of mega-fauna	Human occupation of much of continent Fire stick 'farming' Management of landscape to increase herbivores Vegetation shift to fire promoting species Human population limited by ecological constraints	Development of stone tools Introduction of dingo Sustainable resource management continued Mainland large fauna extinctions (e.g. thylacine)	European settlement and expansion Aboriginal displacement Cessation of Aboriginal land management Sheep and cattle grazing Land clearing for cropping, grazing and construction Fauna declines and extinctions	Soldier settlement Increased mechanisation Increased cropping Increase in parks and reserves Large scale habitat loss and degradation Soil erosion Feral plants and animals and impacts on native species	Vast scale of land clearing Agricultural intensification Irrigated crops and pastures Massive biodiversity loss Rise of environmental movement and Landcare Increasd private land conservation

Figure 3.1: Chronological sequences of agricultural development and environmental impacts and responses from Britain and Australia (not to chronological scale). British timeline based upon information from Rackham (1986) and Simmons (2001); Australian timeline based upon information from Flannery (1994), Roberts *et al.* (2001), Dodson and Mooney (2002), Johnson and Wroe (2003).

These indicate that land use change can be manipulated to increase biodiversity values in agricultural landscapes, and provides some optimism for future conservation efforts.

2. The history of agricultural development within a region can have a significant effect on biodiversity

Land use change scenarios and trajectories differ greatly around the world. For instance, the clearing of native vegetation is generally more prevalent at lower latitudes and in the southern hemisphere, while the intensification of production land is prevalent at higher latitudes in the northern hemisphere (Attwood *et al.* 2009). Within these broad generalisations, there are some well-known biodiversity responses – e.g. clearing of native vegetation tends to elicit a decline in species that are highly specialised to those habitat types (e.g. frugivores, large predators), and an increase in generalists, introduced species and 'edge habitat' or open-country specialists (Marvier *et al.* 2004). Management intensification of land already heavily cleared often leads to a decrease in species that have become adapted to early-successional habitats that are maintained by traditional agricultural management (*sensu* Sutherland 2004).

Inter-regional differences can be better conceived when directly comparing agricultural development and environmental response timelines between markedly differing regions (see Figure 3.1), in this case, Great Britain and Australia. Britain is typified by early historical clearance of native vegetation, a long historical association with low input agriculture and a biota that is adapted to, or has followed the geographical advance of, extensively farmed habitats (Donald *et al*. 2002). Recent biodiversity losses are associated with the rapid intensification of agricultural practices and land use. Australia by contrast has a history of Aboriginal management lasting many millennia, which was interrupted by the arrival of Europeans in the 18th Century. This rapid transition from a system consisting entirely of native vegetation to one dominated in many areas by European-style agriculture has wrought enormous changes on native ecosystems, including massive loss and modification of native vegetation, species extinctions and ongoing species declines and potentially high extinction debt.

3. Land-use history greatly influences regional or national conservation priorities and approaches – in Australia and elsewhere. There is a need to better combine conservation in native vegetation and conservation in the agricultural matrix

The regional differences in land-use change history and trajectories, coupled with the differences in biotic response have led to very different priorities and approaches to biodiversity conservation in different regions, and this is very much reflected in private land biodiversity incentive schemes. Staying with the British-Australian comparison, British agri-environment schemes generally focus on production land (e.g. reversion of arable land to grassland) and semi-natural features (e.g. hedgerows). However, there has been criticism of this focus and suggestions that more broad scale 'climax' habitat restoration should take place (Sutherland 2004), as well as numerous 're-wilding' projects across the country. By contrast, equivalent Australian schemes generally focus biodiversity conservation efforts on native vegetation, often that which has been preferentially cleared and listed as threatened. Until recently little activity has taken place in the agricultural matrix explicitly for biodiversity conservation outcomes. However, there is a growing body of research from Australia and other 'frontier' agricultural regions globally (regions where modern agriculture is a relatively recent introduction) which indicates that a) the agricultural matrix exerts strong influences (e.g. through edge effects, isolation) on native vegetation remnants (Gascon *et al*. 1999), b) contains numerous habitat features, such as scattered 'paddock' trees (Manning *et al*. 2006), and c) provides opportunities for better integration of agricultural production and biodiversity conservation (Attwood *et al*. 2009). Consequently, and with rising concerns about the impact on Australian biota not just of native vegetation loss and degradation but also of the intensification of agricultural land use (Maron and Fitzsimons 2007), there have been recent moves to better integrate conservation in native remnants and conservation in the agricultural matrix.

4. Conservation initiatives should be informed by good science, but there are systemic obstacles to policy–research integration

Globally, it is becoming apparent that reserve networks alone are not sufficient to conserve biodiversity in the long term (Rodrigues *et al*. 2004). This is coupled with growing recognition of the role that private land and private land managers can play in shaping conservation outcomes. Consequently, incentive schemes aimed at enabling land managers to conduct conservation management on private land are an increasingly popular means of achieving conservation outcomes in many countries. Such schemes invest large sums of money in conservation actions, often over relatively long time periods. There is also a growing expectation that policy and programs be founded on an evidence basis (Sutherland *et al.* 2004), and increasing calls for better integration and communication between scientists and decision makers (Gibbons *et al.* 2008). Thus, developing a more collaborative approach between scientists and decision makers in relation to incentive schemes may result in more effective scheme design, improved targeting of investments, better conservation outcomes, and a greatly increased knowledge base of how best to conduct conservation on private land or conservation more generally.

Conservation incentive schemes do provide a potentially fertile collaboration vehicle for those working in science, policy and programs. However, there is a well-recognised systemic disjunct between the spheres of science and policy, which often inhibits interaction. A number of potential explanations have been postulated, including: the academic publication system providing little incentive for researchers to collaborate with policy makers; policy makers and program officers not necessarily having much incentive to engage with the science; policy often requiring generalisations whereas ecological science is often complex and location specific; and increased politicisation in many government departments, where decisions are made on priorities other than science (Gibbons *et al.* 2008; Cummings *et al.* 2010). Potential solutions to this lack of interaction have been postulated by the same authors, including scientists gaining an improved understanding of: (1) policy makers' bureaucratic and hierarchical system, (2) the public service structure to reward scientific literacy, (3) how to build greater interaction/relationship building, and (4) the creation of dedicated 'go-between' knowledge broker positions to act as a conduit and translator between the two spheres.

Other potential impediments to the adoption of a scientifically robust approach to policy and programs include the often short timeframes and rigid processes within which the public service operates, the frequent requirement to expend funds within a financial year (often inhibiting opportunities for long-term planning), and budget profiles that may not reflect a need for an evidence base. Finally, in some instances, the scientific information required by policy and program officers may simply not yet be available; this latter point indicates that policy and program knowledge requirements should be considered in setting research priorities.

5. Scientifically underpinned conservation incentive programs provide an opportunity for better integration and collaboration between researchers, policy makers and program officers. This can lead to a better knowledge base for all and improved conservation outcomes

Recently, the Australian Government has collaborated with scientists and recognised experts to develop an evidence-based approach to the design and delivery of a biodiversity conservation incentive program. The Environmental Stewardship Program (ESP) is a market-based program targeting matters of National Environmental Significance (e.g. threatened ecological communities), with the objective of maintaining and/or improving condition and extent of environmental assets on private land (Commonwealth of Australia 2009). Three areas where ESP has endeavoured to utilise the best available science to inform design and delivery are:

- Metric design and on-ground project evaluation – The ESP is a competitive program, with all bids to participate subjected to a rigorous scoring and evaluation process in order to determine conservation-value-for-money. A conservation value measure tool (CVM tool) is used to evaluate a range of quantitative ecological and other characteristics for each bid. Ways in which science and expert knowledge have recently contributed to the ongoing development of the CVM tool include:
  - Adopting an adaptive approach, with CVM tool performance being frequently reviewed and the findings being used to improve on future iterations of the tool
  - development of state and transition models for target ecological communities that describe the impact of intensified land use upon ecological condition. The models have been used to score sites and communicate ecological condition, threats and management
  - expert-derived probability matrices that estimate the severity of threats impacting a site and the probability of management actions mitigating threats
  - inclusion of 'beyond-patch' management actions intended to a) buffer the 'patch' from external threats from intensive land use and b) reduce the threat of 'patch' isolation by managing connectivity.
- Development of on-ground actions – To deliver on the Australian Government's conservation objectives, it is essential that the management actions proposed are a) appropriate for the ecological community, b) informed by, and complementary to, existing regulatory requirements, c) suitable for the identified condition state of each 'patch' of the community, d) relevant to identified threatening processes where appropriate, e) able to be scored in the CVM tool, and f) inclusive of actions that can be conducted external to the 'patch' to reduce threats from surrounding intensive land uses. These criteria were established through exhaustive literature reviews, workshops with experts for each ecological community, engaging scientists to construct and field test state and transition models for each community, and utilising available research for developing 'beyond-patch' management actions to address intensive land use threats. (See Sutherland *et al.* (2004) and Cook *et al.* (2009) regarding paucity of evidence basis to on-ground conservation actions in UK and Australia respectively.)
- Long term monitoring – A frequent criticism of many European agri-environmental schemes has been that there has often been little monitoring of outcomes (Klein and Sutherland 2003), and therefore little indication as to scheme effectiveness. To avoid such problems, the Australian Government is working with researchers at The Australian National University who have designed and are implementing an ecological monitoring project for a subset of ESP-managed sites. The aims of the monitoring are to:
  - monitor vegetation condition, composition and structure on all sites over time, in order to determine if ESP management is affecting vegetation

- monitor selected fauna groups (birds, reptiles) to determine if changes to vegetation are being expressed as changes in habitat value
- conduct grazing experiments to determine which livestock grazing regimes are most effective for conservation.

The information derived from these studies will be used to report on the progress of the ESP and provide feedback into future ESP management approaches. Having a scientifically informed structure and content should enable the Program to better conserve its target assets and protect them from intensive land use in agricultural landscapes.

Conclusions

Land-use change has considerable influence on many facets of biodiversity, with the importance of land-use transitions on private land becoming an increasing focus for conservationists. Accordingly, schemes aimed at incentivising biodiversity conservation on private land are an increasingly prevalent means of pursuing conservation outcomes globally. To maximise their chances of success, such schemes need to be well designed and implemented, and informed by robust knowledge where possible. While there are a number of systemic obstacles to the integration of science and policy/program development and delivery, such schemes also provide valuable opportunities for the spheres of research and policy to more effectively interact.

Acknowledgements

We are grateful to all those who have worked on the Environmental Stewardship Program since 2007, and more recently S. Whitten, V. Doerr, E. Doerr and A. Langston of the CSIRO for tirelessly helping to develop the Program and its tools. We also wish to thank C. Zammit for comments on an earlier draft. Finally we would like to thank the organisers for inviting S. Attwood to attend the land use workshop and the opportunity to submit a chapter.

Biographies

Simon Attwood is a Program Officer in the Environmental Stewardship Program. His Ph.D. examined arthropod assemblage responses to agricultural land-use change. He has been involved in a wide range of research and extension work in the UK and Australia relating to biodiversity conservation in production landscapes, ecosystem service values of natural, semi-natural and anthropogenic habitats, and mechanisms for delivering on-farm incentive schemes.

Emma Burns is an Assistant Director in the Environmental Stewardship Program. She has a Ph.D. in conservation genetics and phylogeography. She is responsible for scientific management issues to support the Environmental Stewardship Program, including scientific monitoring, and reverse-auction tender and metric designs for optimal conservation outcomes.

She has worked in various roles with a focus on conservation management and ecology for a number of years, including positions in research, consultancy and government.

References

Attwood SJ, Maron M, House APN and Zammit C (2008) Do arthropod assemblages display globally consistent responses to intensified agricultural land-use and management? *Global Ecology & Biogeography* **17**, 585–599.

Attwood SJ, Park SE, Maron M, Collard SJ, Robinson D, Reardon-Smith KM and Cockfield G (2009) Declining birds in Australian agricultural landscapes may benefit from aspects of the European agri-environment model. *Biological Conservation* **142**, 1981–1991.

Attwood SJ (2010) The impacts of agricultural intensification on arthropod assemblages at global and local scales. Ph.D. Thesis, University of Southern Queensland. http://eprints.usq.edu.au/8854/

Commonwealth of Australia (2008) Commonwealth Environment Research Facilities – Research Program Overview. http://www.environment.gov.au/about/programs/cerf/

Commonwealth of Australia (2009) Environmental Stewardship Strategic Framework. http://www.nrm.gov.au/publications/frameworks/environmental-stewardship.html

Cook CN, Hockings MT and Carter RW (2009) Conservation in the dark? The information used to support management decisions. *Frontiers in Ecology and the Environment* **8**, 181–186.

Cummings J, Peters P, Dovers S, Tasker L and Driscoll DA (2010) Workshop report: 'The Worlds of Ecology and Environmental Policy: Never the Two Shall Meet?' *Ecological Management & Restoration* **11**, 152–156.

Cunningham RB, Lindenmayer DB, Crane M, Michael D, McGregor C, Montague-Drake R and Fischer J (2008) The combined effects of remnant vegetation and tree planting on farmland birds. *Conservation Biology* **22**, 742–752.

de Schutter O (2010) 'Agroecology and the right to food.' Genève, United Nations, Human Rights Council, 21.

Dodson JR and Mooney SD (2002) An assessment of historic human impact on south-eastern Australian environmental systems, using late Holocene rates of environmental change. *Australian Journal of Botany* **50**, 455–464.

Donald PF, Pisano G, Rayment MD and Pain DJ (2002) The Common Agricultural Policy, EU enlargement and the conservation of Europe's farmland birds. *Agriculture, Ecosystems and Environment* **89** 167–182.

Fischer J, Brosi B, Daily GC, Ehrlich PR, Goldman R, Goldstein J, Lindenmayer DB, Manning AD, Mooney HA, Pejchar L, Ranganathan J and Tallis H (2008) Should agricultural policies encourage land sparing or wildlife-friendly farming? *Frontiers in Ecology and the Environment* **6**, 380–385.

Flannery T (1994) *The Future Eaters: An Ecological History of the Australasian Lands and People*. Grove Press, New York.

Gascon C, Lovejoy TE, Bierregaard RO Jr, Malcolm JR, Stouffer PC, Vasconcelos HL, Laurance WF, Zimmerman B, Tocher M and Borges S (1999) Matrix habitat and species richness in tropical forest fragments. *Biological Conservation* **91**, 223–229.

Gibbons P, Zammit C, Youngentob K, Possingham HP, Lindenmayer DB, Bekessy S, Burgman M, Considine M, Felton A, Hobbs RJ, Hurley K, McAlpine C, McCarthy MA, Moore J, Robinson D, Salt D and Wintle B (2008) Some practical suggestions for improving engagement between researchers and policy-makers in natural resource management. *Ecological Management & Restoration* **9**, 182–186.

Johnson C and Wroe S (2003) Causes of extinctions of vertebrates during the Holocene of mainland Australia: arrival of the dingo or human impact? *The Holocene* **13**, 109–116.

Kleijn D and Sutherland WJ (2003) How effective are European agri-environment schemes in conserving and promoting biodiversity? *Journal of Applied Ecology* **40**, 947–969.

Lambin EF, Geist HJ and Lepers E (2003) Dynamics of land use and land cover change in tropical regions. *Annual Review of Environment and Resources* **28**, 205–241.

Lindenmayer DB, Knight EJ, Crane MJ, Montague-Drake R, Michael DR and MacGregor CI (2010) What makes an effective restoration planting for woodland birds? *Biological Conservation* **143**, 289–301.

Manning AD, Lindenmayer DB, Barry SC and Nix HA (2006) Multi-scale site and landscape effects on the vulnerable superb parrot of south-eastern Australia during the breeding season. *Landscape Ecology* **21**, 1119–1133.

Maron M and Fitzsimons JA (2007) Agricultural intensification and loss of matrix habitat over 23 years in the West Wimmera, south-eastern Australia. *Biological Conservation* **135**, 587–593.

Marvier M, Kareiva P and Neubert MG (2004) Habitat destruction, fragmentation, and disturbance promote invasion by habitat generalists in a multispecies metapopulation. *Risk Analysis* **24**, 869–878.

Rackham O (1986) *The History of the Countryside*. Dent, London.

Roberts, RG, Flannery TF, Ayliffe LK, Yoshida H, Olley JM, Prideaux GJ, Laslett GM, Baynes A, Mith MA, Jones R and Smith BL (2001) New ages for the last Australian megafauna: Continent-wide extinction about 46,000 years ago. *Science* **292**, 1888–1892.

Rodrigues ASL, Andelman SJ, Bakarr MI, Boitani L, Brooks TM, Cowling RM, Fishpool LDC, Fonseca GABD, Gaston KJ, Hoffmann M, Long JS, Marquet PA, Pilgrim JD, Pressey RL, Schipper J, Sechrest W, Stuart SN, Underhill LG and Waller RW (2004) Effectiveness of the global protected area network in representing species diversity. *Nature* **428**, 640–643.

Sala OE, Chapin FS, Armesto JJ, Berlow E, Bloomfield J, Dirzo R, Huber-Sanwald E, Huenneke LF, Jackson A, Kinzig RB, Leemans RD, Lodge M, Mooney HA, Oesterheld M, Poff NL, Sykes MT, Walker BH, Walker M and Wall DH (2000) Global biodiversity scenarios for the year 2100. *Science* **287**, 1770–1774.

Simmons IG (2001) *An Environmental History of Great Britain: From 10,000 Years Ago to the Present.* Edinburgh University Press, Edinburgh.

Sutherland WJ (2004) A blueprint for the countryside. *Ibis* **146**, 230–238.

Sutherland WJ, Pullin AS, Dolman PM and Knight TM (2004) The need for evidence-based conservation. *Trends in Ecology and Evolution* **19**, 305–308.

Zammit C, Attwood SJ and Burns E (2010) Using markets for woodland conservation on private land: lessons from the policy-research interface. In *Temperate Woodland Conservation and Management.* (Eds DB Lindenmayer, A Bennett and R Hobbs) pp. 297–307. CSIRO Publishing, Melbourne.

4 INTENSIVE FARMING AND ITS ROLE IN WILDLIFE CONSERVATION: ROUTES TO SQUARING THE CIRCLE?

Tim G. Benton

Lesson #1. Scales beyond the farm are important for what happens on the farm.

Lesson #2. Sustainability is in the eye of the beholder.

Lesson #3. Farming systems are not randomly distributed.

Lesson #4. Sustainable agricultural landscapes do not necessarily require 'sustainable' farms.

Lesson #5. Extensive and intensive farming may each be the best option for food and wildlife depending on the place.

Lesson #6. Sustainable production growth is the solution to global food security.

Introduction: the food security challenge

The world's population is predicted to increase by 35% by 2050 (UNPD 2011); additionally, as individual wealth increases, consumption also increases. These two factors suggest global demand for food will grow at a greater rate than the population and although there are uncertainties, the most widely cited prediction is the FAO estimate that 70% more food will be required by 2050 (Bruinsma 2009). Despite the potential for decreasing post-harvest losses, it is likely that global food production will need to continue increasing at rates similar to those of the last two decades (Foresight 2011). Meeting this demand will be a societal challenge as (1) land is increasingly used for non-food crops such as cotton, oil palm and biofuels; (2) environmental degradation such as soil erosion and salinisation has led to abandonment of agricultural land (Smith *et al.* 2010); (3) climate change will have major impacts on agricultural productivity and practices (Lobell, *et al.* 2008; Battisti and Naylor 2009); and (4) movement towards a low carbon economy, coupled with tighter environmental regulations, suggests that agriculture will need to limit the use of agrochemicals, mitigate green house gas (GHG) emissions, and sequester carbon in soils and biomass. This implies that the historical growth of productivity, which is largely based on energy intensive agricultural inputs, will become more difficult to maintain.

There is space to expand the global land area under agriculture (Fischer *et al.* 2002), but this is necessarily limited. First, some of the potential land is tropical forest, and as deforestation is the second major driver of GHG (Smith *et al.* 2010), using this land in agriculture is counterproductive as it would increase the rate of climate change and, therefore, will require more costly mitigation, while simultaneously impacting on the world's most biodiverse habitat. Second, productive land is typically the first to be used for agriculture, suggesting diminishing returns if cultivation expands into marginal areas. Third, land is used for many other services (from biodiversity conservation to human habitation to tourism to carbon storage) (TEEB 2010), creating strong competition preventing unlimited expansion of the agricultural area.

At the same time as global demand is increasing, there is growing recognition that agriculture needs to become more environmentally 'sustainable' (in the sense that degrading services should not impact on future generations (WCED 1987)). The value of the ecological services provided in agricultural landscapes is only just beginning to be recognised (Costanza *et al.* 1997; TEEB 2010) but there are clear indications that the services ecosystems supply have a direct value in production systems, and may become more important in future agriculture, especially when chemical inputs and mechanisation may be restricted by carbon costs. Clearly ecosystems also provide a range of valued services that do not directly aid production, for example carbon storage and the existence of biodiversity.

The demands for both increased sustainability and increased production create a very real tension: how do we increase production sustainably without increasing the land used? Reducing farming intensity is a necessity, given the move to a low-carbon economy, but yields still need to increase. A wholesale conversion to non-intensive (i.e. extensive) production is not an option given that yields are typically lower, requiring more land to produce the same food: land which isn't available. My list of 'lessons learned' comes from a personal academic journey to try and square the circle of conserving as much as we can while producing the food that the global population needs.

Lessons

1. Scales beyond the farm are important for what happens on the farm

We are increasingly recognising that the scale of management for conservation is often greater than the scale at which an individual farmer manages their land. So, we need to consider multiple spatial scales.

All organisms are mobile at some stage of their life cycle; and as a result, in any given place, there is interchange between the immediate habitat and the species pool in the surrounding landscape. As a result, ecosystems, and the services they provide in an agricultural region, reflect not only the organisms present in the cultivated fields but also those in the landscape around the agricultural land (Weibull *et al.* 2003; Gabriel *et al.* 2006; Batáry *et al.* 2010; Gabriel *et al.* 2010), and the persistence of any population depends on having suitably connected habitat across large spatial scales. It therefore makes little sense to think about wildlife conservation simply at the scale of any single management intervention (such as designing a conservation margin prescription for a field): even the best margin will not provide ecosystem services and biodiversity if it is a small isolated patch in an inhospitable matrix. Therefore, management of ecosystem services requires consideration of the field, farm and landscape together.

2. Sustainability is in the eye of the beholder

The definition of 'sustainable' needs to take into account scale effects because a regime of farm management may create positive environmental benefits in one place and negative ones elsewhere (e.g. where winter feed is produced or slurry disposed of).

There are vexed academic discussions about what 'sustainable' means, and this allows the term to be appropriated by people using it to support their often-entrenched world views. For example, extensive farming is often regarded as more sustainable 'by definition'. What is clear is that on an extensively managed field (e.g. an organically managed one) biodiversity is typically higher and synthetic inputs are typically lower. Does this mean that extensive farming is necessarily more sustainable? The answer is no, as sustainability arises from a totality of impacts, both on the actual farm and off it (one particularly important off-farm impact is discussed below). So, if an organic dairy farmer imports winter feed from South America, the environmental costs of this need to be taken into account (in terms of both the carbon cost of production and transport, and the potential environmental cost of deforestation to make agricultural land for soya production). Similarly, selling food from a farm shop may encourage many more local journeys for people to buy local produce, and the sum of these can have a considerable carbon footprint. Finally, some life cycle assessments have revealed that, although organic farms use fewer synthetic inputs (and so there is a carbon saving), they use more fuel in management (mainly due to increased mechanical weeding) and the overall efficiency (output per unit input) may be similar (Gelfand *et al.* 2010).

3. Farming systems are not randomly distributed

If we are to compare different interventions, or management systems, for their environmental costs and benefits, we need to factor out landscape effects and recognise the bias that may arise from landscapes.

Due to the availability and configuration of habitat, different locations will vary both in their 'background' biodiversity and their potential to produce agricultural products, as evidenced by the ubiquity of landscape effects. Some areas, such as hilly ones, will have low potential productivity because of the constraints of slope, valleys, woods and so on, but also have high heterogeneity and therefore high biodiversity (and perhaps high cultural value for the landscape too); other areas may be highly productive with fertile plains with perhaps lower natural biodiversity and lower cultural value. This suggests that both the best conservation strategy (and the impacts of intensification) will be place-dependent. Furthermore, just as we can perceive that landscapes vary in potential productivity, so can farmers, and they clearly modify their farm management appropriately. One corollary of this is that the propensity to become organic is a place-dependent variable as well (Gabriel *et al.* 2009): where farmers may not have high farm yields, the price premium attached to the products of 'conservation' or 'organic' farming may become more attractive. This means that organic farms are more likely to be found in certain landscapes.

Given there is spatial autocorrelation in farm management, one needs to be especially careful in assessing the wildlife value of different farming methods as landscape and farming methods are often confounded (Gabriel *et al.* 2010). It is likely that at least some of the perceived benefit to organic farming is a hidden landscape effect, and when the landscape effect is tightly controlled, the average increase in biodiversity on UK organic farms is about 12% (averaged across groups).

4. Sustainable agricultural landscapes do not necessarily require 'sustainable' farms

Sustainable landscapes can arise without the need for each farm to be 'sustainable' (in the sense of extensive farming): in productive landscapes, more food and more biodiversity may arise from a mix of wildlife areas and intensive farms rather than a landscape full of extensive farms.

To explore the tension between production and conservation, it is useful to think of agricultural *landscapes* as systems that produce two sorts of products: food (and other economic goods like fuel, fibre etc.) and ecosystem services (which may relate to biodiversity, water, carbon storage or environmental health). In a very simplistic sense, there are two basic land management strategies: land can be farmed extensively over the farmable area thereby producing less food but more ecosystem services on the same land ('land sharing'), or farmed intensively over a smaller area and the remaining land can be 'saved' to be managed exclusively for ecosystem services ('land sparing') (Green *et al.* 2005).

Our recent study in the lowland UK (Gabriel *et al.* 2010; Hodgson *et al.* 2010) indicates that if a landscape needs to produce a set amount of food, the optimal way to maximise ecosystem services and biodiversity will sometimes be via land sharing, but is more likely to be via land sparing in productive systems (see Figure 4.1). This suggests that a sustainable agricultural landscape is one with intensive farming producing high yields across a proportion of the landscape, with other areas being deliberately managed for wildlife. This is because semi-natural land, managed for wildlife, has greater biodiversity than extensively managed fields; and similarly, intensively managed fields produce greater yields. Land sparing is then the 'Henry Ford solution': you get more of both farming and wildlife by specialising within the landscape.

5. Extensive and intensive farming may each be the best option for food and wildlife depending on the place

The design of the optimal landscape in terms of land sparing vs. land sharing will depend on the costs (in terms of change in yield) vs. benefits (in terms of biodiversity or ecosystem processes): in low production, high biodiversity landscapes, land sharing may be best, and vice versa.

As discussed above, we have too frequently ignored the landscape context in which production systems sit, and tried to segregate the world into black and white, instead of recognising the shades of grey. Our recent study (Gabriel *et al.* 2010; Hodgson *et al.* 2010; Gabriel *et al. pers.comm.*; Figure 4.1) showed that in a very tightly controlled comparison, organic yields of winter wheat were less than half of those of intensively farmed fields (whereas the biodiversity increase in organic farms was 12%). There are very few 'landscape controlled'

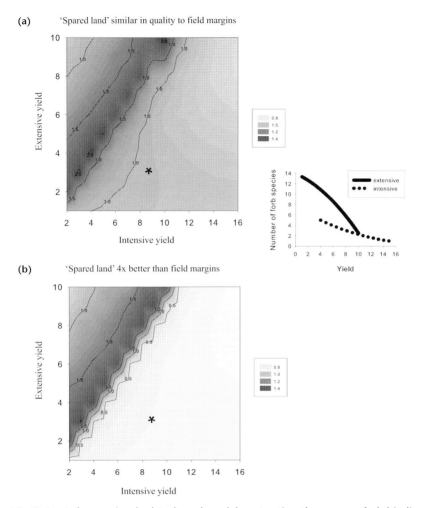

Figure 4.1: Outputs from a simple data-based model contrasting the average forb biodiversity across an arbitrary landscape that is managed either as a land sharing landscape (where there is extensive farming across the totality) vs a land sparing landscape (whereby the same yield landscape is produced in a smaller area of intensive farming, and the excess land is 'spared' for wildlife). The data on forb responses to farming practice are shown in the inset graph (Gabriel *et al.*, unpublished) and the contours show the ratio of biodiversity in sharing scenario vs sparing scenario: when ratio >1 biodiversity, at the same overall landscape yield, is greater in the land sharing scenario. Panel (a) is when the spared land is the quality of margins, panel (b) is when the spared land has 4 times the species richness as margins (as it is for butterflies in nature reserves in our study sites: Hodgson *et al.* 2010). The black asterisk reflects our recorded yields from our matched winter cereal fields differing in whether they were organically or conventionally managed. Where yield in extensive systems is considerably less than in intensive systems, and especially if spared land has higher biodiversity, land sparing may often provide both more food and more biodiversity in the landscape. The model outputs come from a notional landscape of 1000 fields; for each field, yield is derived from a distribution with the given mean and a coefficient of variation as observed (0.79 for extensive, 0.34 for intensive). For that yield, the forb species richness is then derived from the inset graph. For the intensive scenario, the set of highest-production fields that produces the same total yield as the extensive landscape is then kept, and the remainder spared. The average biodiversity across the intensive and extensive landscapes are calculated and the ratio displayed.

like-for-like comparisons, but there is general agreement that extensive yields are lower than intensive yields. So, converting farms in a highly productive area to extensive farming may gain little biodiversity (if the landscape does not support a range of habitats and a large species pool) but will cost much in yield. Conversely, a conventional farm converting to extensive farming in a complex landscape, where production is already constrained by geography and topography, may lose little yield but gain high biodiversity. Agri-environment policy (designed to support production or wildlife, or better how to manage the balance between them appropriately) needs to weigh the costs of lost production versus the gain in biodiversity in each landscape, while recognising that, as demand grows, if everywhere values environmental protection over agricultural production there will not be enough food!

6. Sustainable production growth is the solution to global food security

Given the need for more food, the lack of extra space in which to expand, per unit area productivity needs to grow if demand growth continues. Therefore we need to find ways to increase production sustainably, which will require conservationists to recognise the need for global production and work to minimise its impacts.

The optimal land use strategy, within the land sparing versus land sharing framework, is one where the system maximises ecosystem services while producing the yields that society requires or demands. This implies that if an area converts to extensive farming and yields drop, farming elsewhere will need to intensify (e.g. by converting extensive into intensive, or converting semi-natural land to farmland) to meet the shortfall in demand. For the solution to be rational, the costs and benefits of different land management strategies need to be *assessed across all affected land* in the system, not in each place alone. It is obviously a truism that farmers manage the land at the field and farm scale, but an awareness of the trade-offs at larger spatial scales is required to design the subsidy and legal policy by which farmers are bound.

The land sparing versus land sharing argument is analysed at the level of the system boundaries. This may be at a landscape or even larger scales. For example, if within a country costs and benefits vary regionally, higher productivity in one region will make a larger contribution towards meeting national production needs (and a country's overall demand), allowing other regions to be relatively extensified or spared. This argument naturally extends to the global scale, a scale at which the entire food system is seen to be approaching breaking point. As a case in point, consider organic farming within the EU. Organic farms tend to support local ecology as their farming practices promote landscape heterogeneity, in addition to the reduction in synthetic applications of fertilisers and pesticides. They also produce lower yields. If consumer-demand leads to more organic production within the EU, the total EU production of agricultural products would fall, leaving a shortfall that will necessarily be filled by imports into the EU from regions such as Asia and sub-Saharan Africa. To meet this increased demand from the EU, such areas would need to increase intensification or bring more land into production (at the same time as their own population growth is demanding greater production for local consumption). Furthermore, the EU is heavily regulated relative to other regions, so increasing intensification elsewhere may result in greater environmental damage than it would in Europe.

Additionally, as biodiversity is typically greater in the warmer parts of the world, the environmental damage caused by an expansion of organic farming in Europe may be proportionately greater than the biodiversity protected in Europe. It has recently been estimated that if Europe increased the proportion of its land devoted to organic farming to 20%, then it is likely that >10 M ha, an area equivalent to the size of Portugal, would be needed from the developing world (von Witzke and Noleppa 2010). Hence, European extensification may conserve European environments, but only through the potential export (and amplification) of the environmental costs to other countries in the globe.

Conclusions

As a conservation biologist, I would love to imagine a world of bucolic charm, filled with extensive, wildlife-friendly farming. However, this future would only be possible with radical societal changes in consumption patterns (e.g. the widespread adoption of vegetarian diets), such that this is surely unlikely. Without a large reduction in global demand (let alone the mitigation of the likely growth in demand), the 'organic world' is probably not the most sustainable solution, as there would be a large environmental cost as more land is necessarily brought into agriculture leading to greater deforestation, greater release of carbon into the atmosphere and greater long-term climatic effects.

My overall conclusion is that production of food has polarised into an intensive versus extensive debate (in any of the many variants of this discussion), implying that we can either have food *or* wildlife (or biodiversity or ecosystem services). However, due to the need to move towards a low carbon economy, and greater environmental regulation, intensive agriculture is already becoming 'greener' (with much lower use of synthetic inputs per unit area). It is increasingly apparent that 'conservation' farming practices such as no-till and low-input agriculture can maintain production, increase efficiency and have lower environmental impacts than 'conventional' farming (and perhaps also lower environmental impacts than organic farming, if one accounts for the extra land needed for organic production). Additionally, the further development of precision agriculture, using remote-sensed data producing high-resolution maps of fields to target inputs, can also radically improve efficiency and reduce inputs. Plant-breeding technologies, including new molecular technologies, are also potential partial solutions for maintaining, or increasing, yield in a 'greener' way. Despite organic farming often being seen as a demonstration of good farming practice, especially in the developing world, there is nonetheless widespread acknowledgement that some synthetic inputs can radically improve yields (Vitousek *et al.* 2009), allow creation of biomass for increasing soil carbon as a side effect, and also both reduce poverty and enhance food security.

Thinking of sustainable landscapes, rather than sustainable farms, reduces the dichotomy between extensive and intensive, and therefore reduces the need for the 'production' and the 'conservation' communities to be in battle. As Smith (2010) concludes: global food security will only be possible if productive areas continue being productive (or increase it), while unproductive areas grow productivity. We need to work together to ensure agricultural landscapes are themselves sustainable, producing the necessary food while maximising biodiversity and services; and this requires – given the need for high-production farming – that conservationists do more thinking about how to work with it to maximise the potential benefits for delivering biodiversity and ecosystem services.

Biography

Tim Benton is Professor of Population Ecology at the University of Leeds. His research interests are focused on how populations and ecosystems respond to environmental change, and agro-ecology is the field focus of these interests, coupled with the use of a laboratory model system and a range of theoretical approaches. Sustainable global food security is the current policy focus of his work. Tim also leads the partnership of UK public funders (across government and the research councils) which addresses the research needs in the area of food security. As the Champion for the Global Food Security programme, Tim coordinates workstreams on a range of interesting issues: from resilience in food supplies, nutrition, water and agriculture and waste in the food chain – many of the solutions to these difficult issues require optimising trade-offs between different stakeholders' interests.

References

Batáry P, Matthiesen T and Tscharntke T (2010) Landscape-moderated importance of hedges in conserving farmland bird diversity of organic vs. conventional croplands and grasslands. *Biological Conservation* **143**, 2020–2027.

Battisti DS and Naylor RL (2009) Historical warnings of future food insecurity with unprecedented seasonal heat. *Science* **323**, 240–244.

Bruinsma J (2009) The resource outlook to 2050: by how much do land, water and crop yields need to increase by 2050? *Expert Meeting on How to Feed the World in 2050.* Rome, FAO.

Costanza R, d'Arge R, deGroot R, Farber S, Grasso M, Hannon B, Limburg K. Naeem S, Oneill RV, Paruelo J, Raskin RG, Sutton P and vandenBelt M (1997) The value of the world's ecosystem services and natural capital. *Nature* **387**, 253–260.

Fischer G, van Velthuizen H, Shah M and Nachtergaele F (2002) *Global Agro-ecological Assessment for Agriculture in the 21st Century: Methodology and results.* Laxenburg: IIASA.

Foresight (2011) *Foresight. The Future of Food and Farming, Final Project Report.* The Government Office for Science, London. http://www.bis.gov.uk/Foresight

Gabriel D, Roschewitz I *et al.* (2006) Beta diversity at different spatial scales: plant communities in organic and conventional agriculture. *Ecological Applications* **16**(5), 2011–2021.

Gabriel D, Carver SJ, Durham H, Kunin WE, Palmer RC, Sait SM, Stagl S and Benton TG (2009) The spatial aggregation of organic farming in England and its underlying environmental correlates. *Journal of Applied Ecology* **46**, 323–333.

Gabriel D, Sait SM, Hodgson JA, Schmutz U, Kunin WE and Benton TG (2010) Scale matters: the impact of organic farming on biodiversity at different spatial scales. *Ecology Letters* **13**, 858–869.

Gelfand I, Snapp SS and Robertson GP (2010) Energy efficiency of conventional, organic, and alternative cropping systems for food and fuel at a site in the US Midwest. *Environmental Science & Technology* **44**, 4006–4011.

Green RE, Cornell SJ, Scharlemann JPW and Balmford A (2005) Farming and the fate of wild nature. *Science* **307**, 550–555.

Hodgson JA, Kunin WE, Thomas CD, Benton TG and Gabriel D (2010) Comparing organic farming and land sparing: optimizing yield and butterfly populations at a landscape scale. *Ecology Letters* **13**, 1358–1367.

Lobell DB, Burke MB, Tebaldi C, Mastrandrea MD, Falcon WP and Naylor RL (2008) Prioritizing climate change adaptation needs for food security in 2030. *Science* **319**, 607–610.

Smith P, Gregory PJ, van Vuuren D, Obersteiner M, Havlik P, Rounsevell M, Woods J, Stehfest E and Bellarby J (2010) Competition for land. *Philosophical Transactions of the Royal Society B-Biological Sciences* **365**, 2941–2957.

TEEB (2010) *The Economics of Ecosystems and Biodiversity: Mainstreaming the Economics of Nature: A synthesis of the approach, conclusions and recommendations of TEEB.* UNEP ISBN 978-3-9813410-3-4.

United Nations Population Division (2011) *World Population Prospect: 2010 revision.* United Nations, NY <http://esa.un.org/unpd/wpp/unpp/p2k0data.asp>.

Vitousek PM, Naylor R, Crews T, David MB, Drinkwter LE, Holland E, Johnes PJ, Katzenberger J, Martinelli LA, Matson PA, Nziguheba G, Ojima D, Palm CA, Robertson GP, Sanchez PA, Townsend AR and Zhang FS (2009) Nutrient imbalances: pollution remains response. *Science* **326**, 665–666.

von Witzke H and Noleppa S (2010) 'EU agricultural production and trade: can more efficiency prevent increasing land-grabbing outside of Europe?' Research Report, University of Piacenca <http://www.appgagscience.org.uk/linkedfiles/Final_Report_Opera.pdf>.

Weibull AC, Ostman O and Granqvist A (2003) Species richness in agroecosystems: the effect of landscape, habitat and farm management. *Biodiversity and Conservation* **12**, 1335–1355.

WCED (1987) *World Commission on Environment and Development, WCED, 1987, Our Common Future.* United Nations.

Part B

Specific case studies

5   LAND USE CHANGES IMPERIL SOUTH-EAST ASIAN BIODIVERSITY

Navjot S. Sodhi, Mary Rose C. Posa, Kelvin S.-H. Peh, Lian Pin Koh, Malcolm C.K. Soh, Tien Ming Lee, Janice S.H. Lee, Thomas C. Wanger and Barry W. Brook

Lesson #1. Deforestation likely causes extinctions.

Lesson #2. Species traits explain their vulnerability.

Lesson #3. Primary forests are most valuable for conserving forest species.

Lesson #4. Fragments lose their conservation value over time.

Lesson #5. Forest degradation influences ecosystem processes.

Lesson #6. Protected areas are critical for residual forest biodiversity.

Introduction

South-East Asia (Brunei, Cambodia, Indonesia, Laos, Malaysia, Myanmar, the Philippines, Singapore, Timor-Leste, Thailand, and Vietnam) is arguably the tropical region of greatest conservation concern. The region's wide variety of ecosystems, including lowland and montane rainforests, mangroves, peat swamp forests and limestone karsts, continue to be under pressure from a still-increasing human population (Sodhi *et al.* 2010a). Land use changes in South-East Asia have been extensive, and less than 50% of the original forest cover remains (Sodhi *et al.* 2010a). The region has experienced one of the highest rates of deforestation in the tropics primarily due to agricultural expansion, logging and urbanisation (Sodhi *et al.* 2010a). Disconcertingly, growing global demands for commodities are driving the rapid expansion of oil-palm and paper-and-pulp industries at the expense of South-East Asian lowland dipterocarp forests (Koh and Wilcove 2008). The massive land use changes in South-East Asia are expected to result in concomitant biodiversity attrition. Both literature synthesis and meta-analysis show that species richness and abundance of forest taxa decline from mature to disturbed forests (Sodhi *et al.* 2009; Sodhi *et al.* 2010b). This should be of concern primarily because South-East Asia harbours the highest endemism and endangerment in the tropics for many taxonomic groups (Sodhi *et al.* 2010a).

In this chapter we highlight the major conclusions drawn from studies conducted by the members of Conservation Ecology Laboratory (and its collaborators) at the National

University of Singapore over the past 15 years to understand the effects of land use changes on South-East Asian biotas.

Lessons

1. Deforestation likely causes extinctions

While deforestation is becoming ubiquitous across the tropics, its role in species disappearances is poorly documented. We found that since 1819, over 95% deforestation in densely populated Singapore (540 km^2) has led to more than 28% loss in biodiversity (i.e. vascular plants, arthropods and vertebrates) (Brook *et al.* 2003; Figure 5.1). Extinctions were high (34–43%) in butterflies, freshwater fish, birds and mammals. Factors such as generation time, specific habitat requirements and mobility may have contributed to differing extinction rates among taxa. Indicating the high vulnerability of forest species, extinctions among forest specialists was almost five times higher than among non-forest specialists (Brook *et al.* 2003). Because Singapore is a landbridge island with low levels of species endemism, these biodiversity losses represent local rather than global extinctions. Nevertheless, considering that some plants live for centuries, many of the currently extant species in Singapore could be committed to extinction ('living dead') due to unviable populations. This is illustrated by plant data from an isolated 4 hectare (ha) fragment of lowland rainforest in Singapore (Singapore Botanic Gardens) where half of the tree species

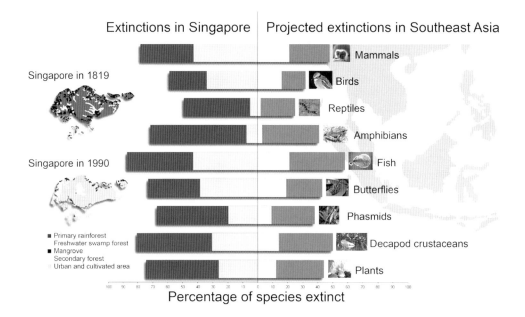

Figure 5.1: Population extinctions in Singapore and South-East Asia. Black and grey bars on the left represent recorded and inferred extinctions in Singapore, respectively. Dark grey and light grey bars on the right represent minimum and maximum projected extinctions in South-East Asia, respectively. Reproduced with permission from Sodhi *et al.* (2004), *Trends in Ecology and Evolution* 19, 654–660.

(>5 cm diameter at breast height) were represented by only one or two individuals, indicating that their populations are unsustainable (Turner *et al.* 1996). It is highly likely that other taxa in Singapore face a similar predicament.

Similarly, a 42 ha island off north Borneo lost 55% of damselflies and dragonflies (Odonata) and 40% of butterflies (Papilionoidea, Hesperioidea) with the complete loss of closed canopy forest between 1928 and 2007 (Sodhi *et al.* 2010c). Data linking deforestation with extinctions are generally not forthcoming. Between the early 1900s and 2008, we found that bird extinctions were fewer on six Malaysian and Indonesian islands with greater remaining forest cover (Sodhi *et al.* 2010d). All these islands were mostly forested in the early 1900s. This study indicates at least a correlation (although not necessarily a causal relationship) between deforestation and extinctions.

2. Species traits explain their vulnerability

Our studies from various taxonomic groups show that large-sized and specialist species are usually more vulnerable to decline due to land use changes. Data from various taxa reveal the high vulnerability of large-bodied animals to extinction (e.g. Brook *et al.* 2003; Lee *et al.* 2009). A common explanation for this trend is that body size is inversely correlated with population size, making large-bodied animals particularly vulnerable to environmental perturbations (Sodhi *et al.* 2004). The extinction proneness of large-bodied animals is further exacerbated because of other correlated traits such as large area requirements, greater food intake, high habitat specificity and low reproductive rates (Sodhi *et al.* 2004).

In Singapore, 76% of insectivorous birds disappeared between 1923 and 1998, likely because of deforestation (Castelletta *et al.* 2000). Insectivores may be more extinction-prone than other feeding guilds because of the impoverishment of insect fauna coupled with their inherently poor dispersal abilities. We also found restricted-range bird species to be susceptible to habitat disturbance (Posa and Sodhi 2006; Soh *et al.* 2006; Sodhi *et al.* 2010b). Species with small geographic ranges are usually specialised and have low local abundance. Therefore, small ranges may make species more vulnerable to habitat perturbations even if local abundance is high.

The most extinction-prone of plant species were likely to have the following traits: epiphytic, monoecious, hermaphroditic, forest dependent and reliant on mammal pollinators (Sodhi *et al.* 2008). The underlying mechanisms explaining the vulnerability of plants are not clear but the explanation may, at least partially, include habitat specialisation and loss of affiliates (e.g. pollinators).

In the case of dung beetles (Scarabaeinae, Aphodiinae), larger mean body lengths were found in old-growth compared to disturbed forest habitats (Lee *et al.* 2009). Increasing disturbance in forest habitats was correlated with an increase in ambient temperature and decrease in ground and shrub cover (Lee 2009). The resulting hotter and drier conditions in human-modified forests could be inhospitable to large-bodied dung beetles, which tend to dissipate heat slower.

3. Primary forests are most valuable for conserving forest species

Our bird studies from Indonesia, Peninsular Malaysia and the Philippines show that old-growth forests contained the highest richness and abundance of forest birds compared to disturbed forests (Peh *et al.* 2005; Sodhi *et al.* 2005a; Posa and Sodhi 2006; Soh *et al.* 2006).

This was also similar for amphibians, but not reptiles in Sulawesi (Indonesia) (Wanger *et al.* 2010). Therefore, the preservation of old-growth forests should be high on conservationists' agenda. Nonetheless, some of the regenerating secondary forests can contain up to two-thirds of forest bird species, and therefore these should be protected to augment old-growth forest reserves. Data on birds and butterflies indicate that conversion of primary or secondary forests into oil-palm plantations can result in significant biodiversity losses (Koh and Wilcove 2008). However, other types of plantations in the vicinity of old-growth forests can be of high conservation value if they are sustainably managed with a limited use of pesticides (Tscharntke *et al.* 2011).

4. Fragments lose their conservation value over time

A large proportion of tropical forests are now fragmented. However, few fragments have been monitored for more than 50 years, and thus their conservation value over the long term is poorly known. The avifaunal changes over 100 years (1898–1998) in a 4 ha patch of rainforest in Singapore (Singapore Botanic Gardens) revealed that 49% of forest species were lost during this time. By 1998, 20% of the birds observed in this site were introduced species (e.g. the House Crow, *Corvus splendens*; Sodhi *et al.* 2005b). Similarly, we studied avifaunal turnover in an 86 ha tropical woodlot (Bogor Botanical Gardens, Indonesia) containing 54% native and 46% introduced plant species and receiving a mean of over 80 000 human visitors per month (Sodhi *et al.* 2006). Since its isolation in 1936, subsequent surveys have shown a gradual reduction in avifaunal species richness. By 2004, the original richness of this woodlot declined by 59% (97 to 40 species) and its forest-dependent avifauna declined by 60% (30 to 12 species). All seven forest-dependent bird species that attempted to re-colonise this woodlot by 1987 perished thereafter. Both these studies show that the conservation value of small fragments for forest birds declines over time, and that they may remain vulnerable to competition or predation by invasive species.

5. Forest degradation influences ecosystem processes

Forests play a key role in supporting human wellbeing, not only because they are a vital source of food and water, but also because they can likely mitigate environmental challenges such as floods (Bradshaw *et al.* 2007). However, the impacts of forest disturbance on beneficial ecosystem processes have been poorly studied in South-East Asia. Dung beetles are crucial members of the forest community because they assist in enhancing soil aeration and nutrition, secondary seed dispersal and mitigating the spread of some vertebrate disease vectors. We found that dung removal was generally greater in old-growth forests (up to 70% reduction in dung mass) than in disturbed forests in Peninsular Malaysia and Singapore (Lee *et al.* 2009). This can be attributed to the absence of large tunneller species (e.g. *Copris* sp.) that have been shown to contribute up to 75% of dung removal function in tropical dipterocarp forests (Slade *et al.* 2007). Therefore, decline in species richness and abundance of dung beetles may result in lower dung removal activity, which has subsequent implications for forest regeneration through compromised soil quality and secondary dispersal of seeds. However, research on the effects of human disturbance on ecosystem services is lacking and we urge that more studies be carried out in South-East Asia.

6. Protected areas are important for residual forest biodiversity

How important are forested protected areas (PAs) for native species? To address this question we surveyed eight PAs and their surrounding buffer areas (within 2 km radius) in Sulawesi. We found that forested PAs in Sulawesi contained a higher number of forest and endemic forest bird species than did buffer areas (Lee *et al.* 2007). Many of these species had significantly higher abundance inside PAs than outside (Lee *et al.* 2008). These results suggest that PAs may be critical for the long-term survival of many endemic forest species on this island. Supporting this, we projected that if all existing forested PAs were to disappear in Singapore, most of the residual forest species could vanish (Brook *et al.* 2003). In some disturbed areas, as much as 80% of canopy cover may need to be retained to preserve all forest bird species (Peh *et al.* 2006; Posa and Sodhi 2006; Soh *et al.* 2006), further suggesting that forest reserves may even be critical in human-dominated landscapes to enhance their attractiveness to forest fauna.

Conclusions

The overarching result of our research has been that land use conversion in South-East Asia threatens vulnerable forest species. Biodiversity declines due to degradation from logging and fragmentation of habitats are already apparent. If the conversion of primary forests into agricultural land (especially large-scale monoculture plantations) continues, it is likely to result in species extinctions. Additionally, some ecosystem processes may be compromised by habitat disturbance, but more research on environmental and socio-economic risks associated with the loss of forests is needed. Protecting old-growth forests will be critical for maintaining residual forest biodiversity; however, other conservation approaches such as preserving regenerated forest should also proceed simultaneously. In particular, informed strategies to manage for the mitigation of biodiversity loss in mixed-use landscapes are largely missing. Such measures are urgently needed if we are to conserve South-East Asia's unique flora and fauna.

Acknowledgements

Our research was funded by the National University of Singapore, National Geographic and Lady McNeice.

Biographies

The late **Navjot S. Sodhi** was Professor of Conservation Ecology at the National University of Singapore. His studies were on the effects of habitat degradation on biotas. Among other things, he was well known for his idiosyncratic infectious laugh. He will be missed.

Mary Rose C. Posa is an Instructor at the National University of Singapore.

Kelvin S.-H. Peh is an AXA Post-doctoral Fellow at St. John's College, Cambridge, UK.

Lian Pin Koh is SNF-Professor of Applied Ecology and Conservation at the ETH Zurich. He studies the implications and tradeoffs of alternative land-use and development options in the developing tropics.

Malcolm C.K. Soh is a Senior Teacher at the NUS (National University of Singapore) High School.

Tien Ming Lee is a post-doctoral research scientist at the Earth Institute, Columbia University, USA. He studies the impacts of global environmental change and the implications of human attitudes and behaviours on biodiversity and protected areas.

Janice S.H. Lee is a PhD candidate at the Department of Environmental Sciences in ETH Zurich. Her current research focuses on sustainability challenges of smallholder oil-palm production systems and its impacts on land cover in Indonesia.

Thomas C. Wanger – a Visiting Fellow at Stanford University, USA and PostDoc at the University of Lüneburg, Germany – is interested in pesticide impacts on tropical ecosystems and ecosystem services provided by mammals and amphibians on a global scale.

Barry W. Brook is professor of climate science at the University of Adelaide's Environment Institute. He models the impact of climate change and other global stressors on extinction risk.

References

Bradshaw CJA, Sodhi NS, Peh KS-H and Brook BW (2007) Global evidence that deforestation amplifies flood risk and severity in the developing world. *Global Change Biology* **13**, 1–17.

Brook BW, Sodhi NS and Ng PKL (2003) Catastrophic extinctions follow deforestation in Singapore. *Nature* **424**, 420–423.

Castelletta M, Sodhi NS and Subaraj R (2000) Heavy extinctions of forest-dependent avifauna in Singapore: lessons for biodiversity conservation in Southeast Asia. *Conservation Biology* **14**, 1870–1880.

Koh LP and Wilcove DS (2008) Is oil palm agriculture really destroying tropical biodiversity? *Conservation Letters* **1**, 60–64.

Lee JSH (2009) Species richness and ecosystem functioning of Southeast Asian dung beetle fauna. Masters Thesis. Department of Biological Sciences, National University of Singapore.

Lee JSH, Lee IQW, Lim SL-H, Huijbregts J and Sodhi NS (2009) Changes in dung beetle communities and associated dung removal services along a gradient of tropical forest disturbance in South-East Asia. *Journal of Tropical Ecology* **25**, 677–680.

Lee TM, Sodhi NS and Prawiradilaga DM (2007) The importance of protected areas for the forest and endemic avifauna of Sulawesi (Indonesia). *Ecological Applications* **17**, 1727–1741.

Lee TM, Sodhi NS and Prawiradilaga DM (2008) Birds, local people and protected areas in Sulawesi, Indonesia. In *Biodiversity and Human Livelihoods in Protected Areas*. (Eds NS Sodhi, G Acciaioli, M Erb and A K-J Tan) pp. 78–94. Cambridge University Press, Cambridge, UK.

Peh KS-H, De Jong J, Sodhi NS, Lim SL-H and Yap CA-M (2005) Lowland rainforest avifauna and human disturbance: persistence of primary forest birds in selectively logged forests and mixed-rural habitats of southern Peninsular Malaysia. *Biological Conservation* **123**, 489–505.

Peh KS-H, Sodhi NS, De Jong J, Sekercioglu CH, Yap CA-M and Lim SL-H (2006) Conservation value of degraded habitats for forest birds in southern Peninsular Malaysia. *Diversity and Distributions* **12**, 572–581.

Posa MRC and Sodhi NS (2006) Effects of anthropogenic land use on forest birds and butterflies in Subic Bay, Philippines. *Biological Conservation* **129**, 256–270.

Slade EM, Mann DJ, Villanueva JF and Lewis OT (2007) Experimental evidence for the effects of dung beetle functional group and composition on ecosystem function in a tropical forest. *Journal of Animal Ecology* **76**, 1094–1104.

Sodhi NS, Liow LH and Bazzaz FA (2004) Avian extinctions from tropical and subtropical forests. *Annual Review of Ecology, Evolution and Systematics* **35**, 323–345.

Sodhi NS, Koh LP, Prawiradilaga DM, Darjono, Tinulele I, Putra DP and Tan THT (2005a) Land use and conservation value for forest birds in central Sulawesi (Indonesia). *Biological Conservation* **122**, 547–558.

Sodhi NS, Lee TM, Koh LP and Dunn RR (2005b) A century of avifaunal turnover in a small tropical rainforest fragment. *Animal Conservation* **8**, 217–222.

Sodhi NS, Lee TM, Koh LP and Prawiradilaga DM (2006) Long-term avifaunal impoverishment in an isolated tropical woodlot. *Conservation Biology* **20**, 772–779.

Sodhi NS, Koh LP, Peh KS-H, Tan HTW, Chazdon RL, Corlett RT, Lee TM, Colwell RK, Brook BW, Sekercioglu CH and Bradshaw CJA (2008) Correlates of extinction proneness in tropical angiosperms. *Diversity and Distributions* **14**, 1–10.

Sodhi NS, Lee TM, Koh LP and Brook BW (2009) A meta-analysis of the impact of anthropogenic forest disturbance on Southeast Asia's biotas. *Biotropica* **41**, 103–109.

Sodhi NS, Posa MRC, Lee TM, Bickford D, Koh LP and Brook BW (2010a) The state and conservation of Southeast Asian biodiversity. *Biodiversity and Conservation* **19**, 317–328.

Sodhi NS, Koh LP, Clements R, Wanger TC, Hill JK, Hamer KC, Clough Y, Tscharntke T, Posa MRC and Lee TM (2010b) Conserving Southeast Asian forest biodiversity in human-modified landscapes. *Biological Conservation* **143**, 2375–2384.

Sodhi NS, Wilcove D, Subaraj R, Yong DL, Lee TM, Bernard H and Lim SLH (2010c) Insect extinctions on a small denuded Bornean island. *Biodiversity and Conservation* **19**, 485–490.

Sodhi NS, Wilcove DS, Lee TM, Sekercioglu CH, Subaraj R, Bernard H, Yong DL, Lim SLH, Prawiradilaga DM and Brook BW (2010d) Deforestation and avian extinction on tropical landbridge islands. *Conservation Biology* **24**, 1290–1298.

Soh MCK, Sodhi NS and Lim SLH (2006) High sensitivity of montane bird communities to habitat disturbance in Peninsular Malaysia. *Biological Conservation* **129**, 149–166.

Tscharntke T, Clough Y, Bhagwat S, Buchori D, Faust H, Hertel D, Hölscher D, Juhrbandt J, Kessler M, Perfecto I, Scherber C, Schroth G, Veldkamp E and Wanger TC (2011) Multifunctional shade-tree management in tropical agroforestry landscapes – a review. *Journal of Applied Ecology* **48**, 619–629.

Turner IM, Chua KS, Ong JSY, Soong BC and Tan HTW (1996) A century of plant species loss from an isolated fragment of lowland tropical rain forest. *Conservation Biology* **10**, 1229–1244.

Wanger TC, Iskandar DT, Motzke I, Brook BW, Sodhi NS, Clough Y and Tscharntke T (2010) Land-use change affects community composition of tropical amphibians and reptiles in Sulawesi (Indonesia). *Conservation Biology* **24**, 795–802.

6 HOW AGRICULTURAL INTENSIFICATION THREATENS TEMPERATE GRASSY WOODLANDS

Sue McIntyre

Lesson #1. The evolutionary history of an ecosystem is strongly linked to the way intensive land use affects biodiversity in that ecosystem.

Lesson #2. In an ecosystem limited by nutrients, agricultural intensification is the most severe form of habitat clearance.

Lesson #3. The leakiness of resources from intensive land uses has both positive and negative effects on biodiversity and ecosystem integrity.

Lesson #4. The maintenance of landscape connectivity and woodland health requires a limit to the area of land subject to intensification.

Lesson #5. Intensive land uses lead to the greatest conflicts with biodiversity conservation as allocation of biophysical capital is a constant sum game.

Introduction

Much attention in Australia has been given to the effects of tree clearing and grazing on native fauna and flora, and we have been slower in recognising the main irreversible factor in the destruction of native biodiversity – intensive agriculture. This is probably because the most widely used early methods of farm establishment were the cutting of trees and the release of livestock, which in many ecosystems do not irreversibly modify the basic processes which support native plants and animals. In Australia, intensification came with the use of machinery and the development of artificial fertilisers, allowing large areas of land to be cultivated, fertilised and, in some cases, irrigated. The southern grassy eucalypt woodlands of eastern Australia have been largely cleared and highly modified by these practices and much of the following chapter draws from experience in these ecosystems in temperate and subtropical regions.

While intensification can be usefully described in terms of the amount and types of specific management actions, a more generic way of describing intensification is through the concept of human appropriation of net primary production (termed HANPP, Erb *et al.*

2009). When ecosystem productivity is diverted to human uses, less biophysical capital remains to support non-production species or critical ecosystem functions. In Australian eucalypt grassy woodlands, trees represent the majority of the above-ground standing biomass. A typical situation would be 60–80 t.ha^{-1} for trees (Burrows *et al.* 2000) and 1–2 t.ha^{-1} pasture biomass (Allan and Bell 1996). The loss of trees represents loss of habitat for most fauna as well as loss of some ecosystem functions relating to water and nutrient flows. This model implies a very strong trade-off relationship between conservation of biodiversity and production, which is consistent with my experience and research observations in the agricultural landscapes of southern Australia (McIntyre *et al.* 2002a,b; Dorrough *et al.* 2006; Martin and McIntyre 2007).

Lessons

1. The evolutionary history of an ecosystem is strongly linked to the way intensive land use affects biodiversity in that ecosystem

It is impossible to predict the effects of land use intensification without having some understanding of the selection pressures that have operated on a species assemblage over evolutionary time frames. Contrasting examples from Australian ecosystems provide an illustration of this. Although ecosystem collapse in the face of intensive agriculture is a reasonable generalisation for the majority of Australian ecosystems, there are exceptions. One of these is the flora of intermittent wetlands on the semi-arid Riverine Plain in New South Wales (McIntyre and Barrett 1985). In the face of irrigation development and rice cultivation, native wetland plants persisted in the flooded rice paddies, in sufficient numbers to be considered major economic weeds. A parallel situation occurs in the Californian rice fields (McIntyre and Barrett 1985). In both cases, the native wetland species appear to have been pre-adapted to rice cropping due to exposure to high nutrients (alluvial soils, and flooding make phosphorus more available) and frequent unpredictable disturbance (variable rainfall, rapid drying of shallow wetlands) over evolutionary time frames. In contrast, terrestrial cereal crops are dominated by exotic weed species (Velthuis and Amor 1983) introduced from countries where cropping has been practised for millennia and soils are naturally fertile. The success of exotics over native species is the more common response to intensification observed for woodland and grassland ecosystems in Australia (Dorrough *et al.* 2006; McIntyre 2008).

2. In an ecosystem limited by nutrients, agricultural intensification is the most severe form of habitat clearance

In part due to the aged and weathered nature of Australian landscapes, Australian soils are generally low in nutrients and have evolved a biota adapted to oligotrophic environments (Hopper 2009). Grazing unfertilised grassy vegetation without cultivation or inputs creates a degree of modification, with the loss of grazing sensitive species and some vulnerability to soil erosion due to the harvesting of biomass. However, critical ecosystem functions and plant diversity, together with tree populations, tend to be maintained in a rangeland setting (McIntyre *et al.* 2002b). With intensification, the productivity of the ground layer is

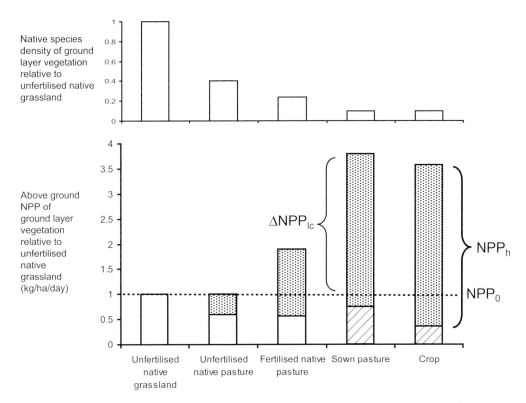

Figure 6.1: Relative changes in native species density and approximate net primary production (NPP) under land uses of increasing intensity for temperate grassy vegetation. NPP is the sum of monthly pasture growth rate under average growing conditions. NPP_0 = potential productivity of the original ecosystem, NPP_h = biomass harvest (shaded area), ΔNPP_{lc} = the productivity change with land use change, hatched area indicates biomass dominated by exotic species. Sources: species density from Dorrough *at al.* (2006); NPP from Appendix 3, Allan and Bell (1996); 40% native pasture harvest rate from Partridge (2000); other utilisation rates estimated by author to be 60, 80, 90% respectively; NPP components Erb *et al.* (2009).

increased through fertilisation, sowing of legumes and cultivation ($DNPP_{lc}$, Figure 6.1) and a large proportion of this is harvested (NPP_h), although with good agricultural practices, a significant fraction of the original productivity (NPP_0) is retained. The significant difference is that in sown pastures and crops, the remaining biomass is dominated by exotics (McIntyre 2008; Figure 6.1) and native species richness is very low (Dorrough *et al.* 2006). While a few native weedy species are able to exploit intensive land uses, the majority do not persist (Dorrough and Scroggie 2008). Intensive agriculture is therefore a highly transforming land use, and cropping and improved pastures effectively replace the original ecosystems and their associated plant species in temperate grassy woodlands.

3. The leakiness of resources from intensive land uses has both positive and negative effects on biodiversity and ecosystem integrity

The aim of intensive farming practices is to provide an environment suitable for the establishment and maximum growth of productive plants. The provision of ample resources in a non-competitive environment for domesticated plants and animals creates an inherently leaky system, and losses of soil, water and nutrients from the production system are typical. Most obviously, leakiness manifests as eutrophication of waterways and run-on onto non-production land with the undesirable effects of algal blooms in aquatic systems and weed incursions in native vegetation. Managers seeking production efficiencies strive to avoid these losses and divert all resources to the means of production, that is crops or domestic livestock. The wide occurrence of eucalypt dieback can be seen as a more subtle manifestation of resource leakiness. Eucalypts can be passively, but permanently, eliminated from landscapes through a complex interaction of intensification and livestock grazing, in some cases mediated by leaf-defoliating insects whose larvae benefit from the increased productivity of fertilised pasture roots (Reid and Landsberg 2000).

Native fauna may exploit opportunities created by intensive land uses. Waterbirds, frogs and tortoises use rice fields (McIntyre *et al.* 2011) and terrestrial birds sometimes feed in pastures and crops. Depending on what the birds are eating in crops, they may be viewed neutrally, as pests, or as pest controllers. From a biodiversity conservation perspective, there are questions around the extent to which intensive land uses are providing critical resources, and whether other elements vital to their survival are available in the landscape. For example the maintenance of breeding sites such as tree hollows or natural wetlands may be in conflict with the interests of continuing or increasingly intensive agriculture (McIntyre *et al.* 2011).

4. The maintenance of landscape connectivity and woodland health requires a limit to the area of land subject to intensification

The important question for biodiversity conservation is not whether intensive land use occurs, but the quality of management in controlling damaging leakiness, and the amount and location of intensive land uses. For example, Wylie *et al.* (1993) found significantly more severe native tree dieback in properties with more than 50% improved pasture. In addition, the nature and extent of the complementary land uses are critical to overall landscape health and the ability of native biota to persist and prosper. With this in mind, landscape design principles were developed for eucalypt grassy woodlands (McIntyre *et al.* 2002b) identifying a stratification of land use intensities and the critical elements associated with each land use to address the needs of the majority of native species. The key idea from this exercise that was recognised in conservation practice was the minimum requirement for 30% woodland cover in landscapes, which is a critical threshold for mobile species (Andrén 1994; Pearson *et al.* 1996) assuming that they require trees as habitat. However, the most demanding species are the ones with the most limited mobility such as plants (Dorrough *et al.* 2012) and invertebrates. That is, species that do not necessarily require trees, but which will not persist in crops or fertilised pastures. This group of organisms need much more habitat to retain connectivity, and this can be provided if intensification is limited to 30% of the landscape. Regardless of configuration, 70% habitat extent (comprising native pastures, grazed woodlands and conservation areas) will provide connectivity for even the most movement-restricted organisms (Pearson *et al.* 1996) and is therefore the most robust and conservative of thresholds.

5. Intensive land uses lead to the greatest conflicts with biodiversity conservation as allocation of land use is a constant sum game

To the disappointment of conservation biologists and land managers alike, there is very little evidence of win-win solutions for biodiversity conservation and agricultural production. The evidence strongly points to the existence of trade-offs between increased production and species conservation at the community level. It is true that in low-input systems, commercial grazing can co-occur with high biodiversity in commercial farming situations, but even in these settings there is some compromise: production is limited and sensitive species are affected (McIntyre *et al.* 2003). In theory, there should be a point at which the net returns of extensive agriculture are compensated for by the lower risk of losses of investment in the years of production failure.

The practical reality is that if land capability allows for intensive production, particularly cropping, then this is the land use that is adopted. There is a set amount of productive land in the landscape and using any of it for conservation directly reduces landholder incomes. This results in overdevelopment of the most fertile landscapes, at least from the point of view of biodiversity conservation. The widely promoted idea of intensification within productive parts of the landscape to allow increased land to be conserved for conservation of biodiversity has little meaning in Australian cropping landscapes – nearly all the landscape has been intensified. In less developed landscapes, the land-sparing question has been examined by Dorrough *et al.* (2007) who found that options for future intensification, via fertiliser application, often coincided with areas of high native plant diversity and further intensification would come at a cost to biodiversity. In contrast, improving grazing management across broad scales was considered likely to result in enhanced profitability and could also benefit native vegetation.

Conclusions

A balanced landscape whereby intensive land uses are restricted to the most productive soils and even then limited to the minority of the landscape is an ecologist's ideal that is seen in limited parts of south-eastern Australia, but generally only when land capability limitations exist. Elsewhere, for example the tropical grassy woodlands of northern Australia, technology and economics have combined to constrain intensification and it is possible to see examples of functional, biodiverse and economically viable farming enterprises. However, such examples are merely the product of chance rather than choice, and we are yet to see conscious adoption of extensive agriculture in any significant way in Australia.

A variety of incentive schemes have been initiated to encourage extensification on farms through compensation for lost production. However, these have been exploratory in nature or small in scope and have yet to have impacts at a district or regional scale. We are at a juncture whereby society could choose to support biodiversity conservation through the scaling-up of incentive programs in rural areas, but current signals are indicating overriding concerns with global change and food security. The consequences of this are likely to be scaling-up and expansion of intensive production systems, with a continuing negative outlook for biodiversity and landscape health.

Acknowledgements

This work draws broadly from many research projects, observations and interactions with colleagues. In particular I would like to acknowledge support from CSIRO, Land and Water Australia, Meat and Livestock Australia and the New South Wales Government through its Environmental Trust.

Biography

Sue McIntyre is a Senior Principal Research Scientist at CSIRO Ecosystem Sciences. She has over 30 years' research experience in a range of ecosystems in Australia and Europe. However, her particular focus is on the grassy eucalypt woodlands, and the impacts of farming practices on biodiversity and ecosystem function.

References

Allan CJ and Bell AK (1996) *PROGRAZE Manual*. New South Wales Agriculture and Meat Research Corporation, Orange, NSW.

Andrén H (1994) Effects of habitat fragmentation on birds and mammals in landscapes with different proportions of suitable habitat: a review. *Oikos* **71**, 355–366.

Burrows WH, Hoffmann MB, Compton JF, Back PV and Tait LJ (2000) Allometric relationships and community biomass estimates for some dominant eucalypts in Central Queensland woodlands. *Australian Journal of Botany* **48**, 707–714.

Dorrough J, McIntyre S, Brown G, Stol J and Barrett G (2012) Differential responses of plants, reptiles and birds to grazing management, fertilizer and tree clearing. *Austral Ecology* doi: 10.1111/j.1442-9993.2011.02317.x.

Dorrough J, Moll J and Crosthwaite J (2007) Can intensification of temperate Australian livestock production systems save land for native biodiversity? *Agriculture, Ecosystems and Environment* **121**, 222–232.

Dorrough J, Moxham C, Turner V and Sutter G (2006) Soil phosphorus and tree cover modify the effects of livestock grazing on plant species richness in Australian grassy woodland. *Biological Conservation* **130**, 394–405.

Dorrough J and Scroggie MP (2008) Plant responses to agricultural intensification. *Journal of Applied Ecology* **45**, 1274–1283.

Erb K, Krausmann F, Gaube V, Gingrich S, Bondeau A, Fischer-Kowalski M and Haberl H (2009) Analyzing the global human appropriation of net primary production – processes, trajectories, implications. An introduction. *Ecological Economics* **69**, 250–259.

Hopper SD (2009) OCBIL theory: towards an integrated understanding of the evolution, ecology and conservation of biodiversity on old, climatically buffered, infertile landscapes. *Plant and Soil* **322**, 49–86.

Martin TG and McIntyre S (2007) Impacts of livestock grazing and tree clearing on birds of woodland and riparian habitats. *Conservation Biology* **21**, 504–514.

McIntyre S (2008) The role of plant leaf attributes in linking land use to ecosystem function and values in temperate grassy vegetation. *Agriculture, Ecosystems and Environment* **128**, 251–258.

McIntyre S and Barrett SCH (1985) A comparison of weed communities of rice in Australia and California. *Proceedings of the Ecological Society of Australia* **14**, 237–250.

McIntyre S, Heard KM and Martin TG (2002a) How grassland plants are distributed over five human-created habitats typical of eucalypt woodlands in a variegated landscape. *Pacific Conservation Biology* **7**, 274–285.

McIntyre S, Heard KM and Martin TG (2003) The relative importance of cattle grazing in subtropical grasslands – does it reduce or enhance plant biodiversity? *Journal of Applied Ecology* **40**, 445–457.

McIntyre S, McIvor JG and Heard KM (Eds) (2002b) *Managing and Conserving Grassy Woodlands*. CSIRO Publishing, Melbourne.

McIntyre S, McGinness HM, Gaydon D and Arthur AD (2011) Introducing irrigation efficiencies: prospects for flood-dependent biodiversity in a rice agro-ecosystem. *Environmental Conservation* **38**, 353–365.

Partridge I (2000) *Managing Grazing in the Semi-Arid Woodlands*. Queensland Department of Primary Industries, Brisbane.

Pearson SM, Turner MG, Gardner RH and O'Neill RV (1996) An organism-based perspective of habitat fragmentation. In *Biodiversity in Managed Landscapes: Theory and Practice*. (Eds RC Szaro and DW Johnston) pp. 77–95. Oxford University Press, New York.

Reid N and Landsberg J (2000) Tree decline in agricultural landscapes: what we stand to lose. In *Temperate Eucalypt Woodlands in Australia: Biology, Conservation, Management and Restoration*. (Eds RJ Hobbs and CJ Yates) pp. 127–166. Surrey Beatty & Sons, Chipping Norton, Sydney.

Velthuis RG and Amor RL (1983) Weed survey of cereal crops in south west Victoria. *Australian Weeds* **2**, 50–52.

Wylie FR, Johnston PJM and Eisemann RL (1993) 'A Survey of Native Tree Dieback in Queensland'. Queensland Department of Primary Industries, Forest Research Institute, No. 16.

7 MITIGATING LAND USE INTENSIFICATION IN THE ENDEMIC-RICH HOTSPOTS OF SOUTHERN AFRICA AND WESTERN INDIAN OCEAN

Michael J. Samways

Lesson #1: Recognition of multiple spatial scales is critical.

Lesson #2: Agricultural or vegetation type has a major influence on land use planning options.

Lesson #3: Landscape-scale ecological networks are effective for mitigating impacts of intense land use.

Lesson #4: Restoration can be highly effective, even after intense land use changes, as long as remnant natural populations remain, and invasive alien organisms are controlled or eradicated.

Lesson #5: Effective bioindicators are essential for assessing the extent and intensity of land use changes, and for monitoring restoration activities.

Lesson #6: Mitigation of land use intensification can be synthesised into an all-embracing strategy for future sustainability using an integrated design and management approach.

Lesson #7: Human communities must be included in land use planning from the beginning.

Introduction

Biodiversity loss in global hotspots (Mittermeier *et al.* 2004) is of concern, as endemic species, traits, functional groups and interactions are irreplaceable when locally threatened. Such hotspots are often also geographical areas of intensive land use, as well as being subject to severe invasions from alien organisms, especially alien plants. Indeed, alien plants are often the result of land use intensification.

Effective mitigation measures must be brought in to deal with the multiple pressures, based on sound science and what is practical and encouraging for land managers (Hobbs and Lindenmayer 2007). The process should be that land use intensification is assessed, mitigation measures are put in place and the success of these steps is then measured with suitable indicators. This process is then repeated. Out of this, a synthetic approach with general principles for land use improvement can be developed. This approach also recognises the traditions and needs of the local people. The following are examples from South Africa where the major driver is to be able to export produce to overseas markets that demand that the agricultural practices include biodiversity conservation, maintenance of near-natural ecosystem processes and fair trade employment. Restoration activities in the western Indian Ocean provide further insight.

Lessons

1. Recognition of multiple spatial scales is critical

Spatial scale is important, as different but critical processes take place at different spatial scales. Understanding these different scales is important for assessing pressures on biodiversity and for mitigation of those pressures.

The agricultural landscape must be viewed at multiple spatial scales, as this is important for instigating measures for mitigation. This is because there are issues relating to within-patch, and those relating to the landscape.

Within-patch mitigation involves modifying the landscape, through activities such as organic farming and making agricultural habitats stepping stone habitats for organisms to disperse across the whole landscape (Gaigher and Samways 2010). These connectivity patches are important, even for narrow-range endemics. Such landscape modification must also be integrated with decision processes that involve financial viability, ecological feasibility and social considerations (Hobbs and Lindenmayer 2007).

2. Agricultural or vegetation type has a major influence on land use planning options

Understanding the footprint of particular types of agroforestry patches is important for finding ways to improve biodiversity conservation and maintain ecosystem processes. Different patch types require different approaches.

Plantation forestry can have a severe impact on biodiversity at the local spatial scale of the patch, i.e. the plantation block. Once the trees mature, there is little opportunity to 'soften' the patch for biodiversity conservation and maintenance of ecosystem processes. Mitigation must therefore take place at the spatial scale of the overall landscape. This can be done using corridors, large-scale ecological networks and adjusting the patch perimeter to improve processes such as hydrological function (known as delineation) (Samways *et al.* 2010a).

Associated with this footprint are edge effects (Lindenmayer and Hobbs 2007), where the edge zone of plantation trees is about 30 m into the remnant grassland, which is similar for vineyards and also for clumps of invasive alien trees (Samways and Moore 1991; Magoba and Samways 2012).

3. Landscape-scale ecological networks are effective for mitigating the impacts of intense land use

Ecological networks are effective for mitigating the effects of land transformation, especially from plantation forestry. Incorporation of ecological networks into overall landscape planning has major benefits for local biodiversity.

Corridors and nodes of remnant natural habitat can be developed into large-scale ecological networks (ENs) in an agroforestry landscape (Samways 2007a). These ENs are a major mitigation feature across South African plantation landscapes (Samways *et al.* 2010a). The quality and width of these ENs, which make up about a third of the overall landscape, is critical. These ENs should be of high quality natural habitat and ideally link protected areas or conservancies across the greater landscape. They function at both the ecological and evolutionary scales of processes. Their effective use in biodiversity conservation and the conservation of ecosystem processes involves the mesofilter, i.e. including features and characteristics of the landscape necessary for the survival of species within the ENs. The mesofilter includes marshland, topographic landmarks, fallen logs, vegetation-bare areas and even dung heaps from indigenous megaherbivores. An additional operational layer, over that of the landscape coarse filter and mesofilter, is that of the species: fine filter. Certain species, whether threatened or functionally important, are targeted for specific attention and conservation.

Research to date on ENs has focused on plantation forestry (as it is the most advanced of all agroforestry sectors in mitigation measures for Forest Stewardship Council certification), and it is now being applied to the agricultural landscape, where the production patches themselves can also be modified for future sustainability.

4. Restoration can be highly effective, even after intense land use changes, as long as remnant natural populations remain and invasive alien organisms are controlled or eradicated

Restoration is second-best to maintaining the landscape intact in the first place, and is a very important activity where many endemic species are concerned. It can only be done when there is a residual population of the local organisms to re-populate the overall landscape, and when invasive alien species are eradicated or controlled.

Invasive alien organisms can have a major impact on local biodiversity, with disturbance caused by agriculture often providing an invasion pathway. Appreciating this is vital for effective mitigation measures for the agricultural landscape. Alien trees can shade out indigenous biodiversity and change ecological processes through taking up large amounts of water. The adverse impact of alien trees can be even greater than that of intense agriculture (Magoba and Samways 2012).

Removal of invasive alien trees can lead to major recovery of biodiversity, even for narrow-range endemics (Samways and Sharratt 2010; Samways *et al.* 2011). Improving the landscape for biodiversity and ecosystem restoration is intrinsic to mitigation in agricultural areas. Impacts of alien vegetation are not adverse for all species. In the Comoro Islands, alien trees provide necessary shading for some endemic invertebrates (Samways 2003).

Restoration can be highly effective, even for narrow-range endemics, when sufficient remnant populations are still present. For the full recovery of endemic populations, invasive alien organisms must be controlled or eradicated, while agriculture is organised into small discrete patches for efficient and sustainable production. When the historical vegetation is restored, various animals naturally then colonise these restored areas. Translocations of species are a last resort. This was tested on the Seychelles island of Cousine where highly invasive alien plants, cats and livestock were eradicated and agriculture confined to a small area. Invertebrates recovered quickly, with some highly threatened land birds having to be re-introduced. Sea birds recovered of their own accord, with the Sooty Tern returning after an absence of 30 years (Samways *et al.* 2010b).

5. Effective bioindicators are essential for assessing the extent and intensity of land use changes, and for monitoring restoration activities

We need yardsticks, or bioindicators, to assess the extent to which land use change is affecting biodiversity, and to determine how well restoration activities are progressing. These bioindicators must be easy to use and effective.

Mitigation of land use intensification requires some measure for determining how well the recovery process is progressing. For this, it is important to have appropriate and effective bioindicators. However, because a species, or group of species, whether actual or functional, are sensitive to land use change, it does not mean that they are a good yardstick unless their responses are known and quantified (McGeoch 2007). Furthermore, these bioindicators must be user-friendly. For example, while ants are often seen as good bioindicators, their use in conservation management requires physically intensive and time-consuming sinking and sorting of pitfall traps, which is hardly likely to be done by practical conservation managers. A suite of species needs to be selected that are viable in practical terms for everyday assessment. These are usually highly visible and regularly apparent species with known, quantifiable responses (Simaika and Samways 2011). With appropriate bioindicators, management vis-à-vis design principles for effective conservation can be disentangled (Bazelet and Samways 2011).

Bioindicators must be representative of biodiversity as a whole, otherwise they are an indicator essentially only unto themselves. Using a single, carefully selected taxon, functional group or even a suite of traits, may be suitable for answering a particular ecological or conservation question (Barton *et al.* 2009; Bazelet and Samways 2011). However, the reason for using a range of taxa is that certain single taxa may be sensitive and resilient to a landscape change, such as fire, while others may also be sensitive but not resilient. We need more honest brokering of biodiversity as a whole and not blithely use one taxon and suggest conservation management activities based solely on the results from that one group (Pryke and Samways 2009).

6. Mitigation of land use intensification can be synthesised into an all-embracing strategy for future sustainability using an integrated design and management approach

In South Africa, there is sufficient knowledge to synthesise a set of comprehensive guidelines for optimising agricultural production without compromising biodiversity and ecosystem processes. These are a practical way forward for conservation managers.

Principles for Integrated Biodiversity Conservation

For effective conservation at the landscape, coarse-filter operational level, there are six interrelated principles that are pivotal for success. Five of these are design principles and one a management principle.

Design principle 1: Maintain natural reserves
Reserves should be as large as possible and as many as possible. 'How large' depends on the organisms in question, and the availability of land. It is important for specialist species.

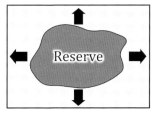

Maximise protected area

Design principle 2: Maintain quality habitat heterogeneity
The aim is to maintain as many natural biotopes as possible at various spatial scales. It is important for maintaining a diversity of opportunities for as many indigenous species as possible. This involves removal or suppression of invasive aliens. The emphasis is on quality, natural heterogeneity, not just any heterogeneity.

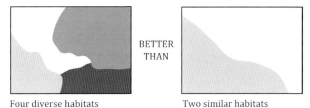

Four diverse habitats   Two similar habitats

Design principle 3: Reduce contrast in vegetation between disturbed areas and adjacent natural areas
Softening hard, contrasting edges allows for more variety of microclimate and micro-habitat conditions, especially for small ectothermic organisms like insects. Soft edges, i.e. gentle ecotones, also improve connectivity within the landscape.

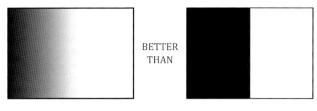

Figure 7.1: An integrated approach to biodiversity conservation involves five design principles, one management principle, and a central tenet. These are all interrelated and applied simultaneously for practical biodiversity conservation and maintenance of ecosystem processes in a production landscape. (Based on Samways *et al.* 2010c).

Design principle 4: Outside reserves, soften the landscape as much as possible
Land sparing or setting-aside of patches is a spatially explicit approach. A more integrated approach between agricultural systems and nature reserves is to instigate conservation headlands, conservancies and other management activities that provide undisturbed or lesser disturbed areas. This improves area of occupancy and connectivity, so increasing chances of survival.

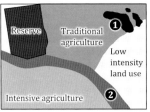

❶ Conservencies may retain patches of undisturbed or less disturbed habitat

❷ Conservation headland with low chemical input, banks of wild flowers and free growing hedgerows

Large, high-quality habitats close together

Design principle 5: Link patches of quality habitat
It is essential to maintain natural ecological conditions and evolutionary potential of landscapes. This allows movement of organisms and maintenance of high population levels to buffer adverse conditions and promote genetic vigour. Full corridors may not always be feasible and stepping stone habitats become a second choice.

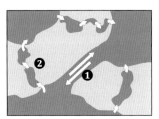

❶ Corridor or linkage of high quality habitat

❷ Stepping stone habitats where corridors are not feasible

Linkages between quality habitats

Management principle: Simulate natural disturbance
Most landscapes experience some disturbance, whether abiotic, such as fire or flooding, or biotic, such as grazing impact from large herbivores. These disturbances may be patchy and at small spatial scales (e.g. trampling around waterholes) or large ones (e.g. extensive grassland fires). The aim is to simulate these natural conditions, with particular attention to extent, intensity and timing of management activity. One mitigation study showed that management was 2-5 times more important than design for maintaining the natural ecological processes in large-scale ecological networks (Bazelet and Samways 2011).

Figure 7.1: Continued

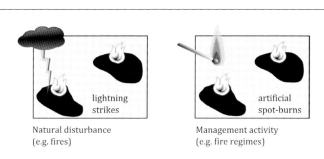

The central tenet: Maintain the trio of large patch size, good patch quality and reduced patch isolation

The six principles above are a landscape level approach. At the population level, it is essential that a large population is maintained through good quality habitat and sufficient habitat size. Habitat is a species' needs and not necessarily a spatially explicit patch. These large, quality habitats also need to be close together to ensure movement between them. This population level approach is the essence of single-species, or fine-filter conservation.

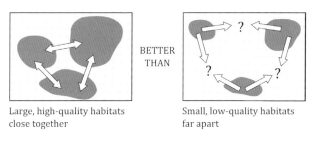

Figure 7.1: Continued

From research and development on biodiversity conservation in production landscapes in biodiversity hotspots, some general principles are beginning to emerge (Samways 2007b). These involve five design principles, one management principle and a central tenet, illustrated in Figure 7.1.

7. Human communities must be included in land use planning from the beginning

Well-designed and managed ecological landscapes must also consider the traditions and needs of human communities. This is important in biodiversity hotspots where there is a long history of agricultural activity.

A production system involving conservation of indigenous biodiversity and maintenance of ecosystem services can be modelled using a Framework for Ecosystem Service Provision (Rounsevell *et al.* 2010). This modelling framework can then be adapted for various socio-ecological systems. First, there are plans for agricultural production alongside conservation mitigation on ecological grounds, and then these plans are adapted according to the needs and traditions of the local community. For example, culturally significant cattle owned by local people graze inside South African ecological networks, and glamorous indigenous flowers are harvested from protected areas in the Cape Floristic Region biodiversity hotspot. However, it is imperative that these local communities are involved in the planning process from its inception to ensure full communication towards a successful outcome.

Conclusions

To mitigate landscape intensification we need sustainable and practical landscape planning and management. This requires a contingency approach: catering for all eventualities, especially adverse weather and climate change. Future production systems need to be robust and resilient. For this, we need a set of researched and tested guidelines that must be easy to implement. We cannot wait for overwhelming scientific certainty, and must apply the Precautionary Principle of insightful conservation management using findings to date. A successful approach has been developed in South Africa, driven by Forest Stewardship Council certification.

   Biodiversity hotspots require careful handling, as risks of losing irreplaceable biota are high. A set of mutually reinforcing landscape level management and design guidelines is now available for practical conservation management across optimally productive agroecosystems. Critical features include recognition of spatial scale in relation to intensity of land use, removal of alien organisms and development of extensive ecological networks. Successful implementation requires a good set of bioindicators and inclusion of all human communities from the start.

Acknowledgements

Great discussions were had with Corinna Bazelet, Chris Burchmore, Peter Gardiner, Lize Joubert, Rembu Magoba, James Pryke and Sven Vrdoljak. Thanks to Linda Broadhurst, Saul Cunningham, David Lindenmayer and Andrew Young for a stimulating workshop.

Biography

Michael Samways is Professor and Chair of the Department of Conservation Ecology & Entomology, Stellenbosch University, South Africa, and is a Fellow of the Royal Society of South Africa and a Member of the Academy of Science of South Africa. He is a John Herschel Medallist of the Royal Society of South Africa, Senior Captain Scott Medallist of the South African Academy of Sciences and Arts and Gold Medallist of the Academy of Science of South Africa. He received the Rector's Award for Excellence in Research in 2006 and 2010 and won the Distinguished Scientist Award at the Global Conference of Entomology in 2011. He has published widely on invertebrate and landscape conservation.

References

Barton PS, Manning AD, Gibb H, Lindenmayer DB and Cunningham SA (2009) Conserving ground-dwelling beetles in an endangered woodland community: multi-scale habitat effects on assemblage diversity. *Biological Conservation* **142**, 1701–1709.

Bazelet CS and Samways MJ (2011) Relative importance of management vs. design for implementation of large-scale ecological networks. *Landscape Ecology* **26**, 341–353.

Gaigher R and Samways MJ (2010) Surface-active arthropods in organic vineyards, integrated vineyards and natural habitat in the Cape Floristic Region. *Journal of Insect Conservation* **14**, 595–605.

Hobbs RJ and Lindenmayer DB (2007) From perspectives to principles: where to from here? In: *Managing and Designing Landscapes for Conservation*. (Eds DB Lindenmayer and RJ Hobbs) pp. 561–568. Wiley-Blackwell, Oxford, UK.

Lindenmayer DB and Hobbs RJ (2007) Synthesis: edge effects. In: *Managing and Designing Landscapes for Conservation*. (Eds DB Lindenmayer and RJ Hobbs) pp. 195–197. Wiley-Blackwell, Oxford, UK.

Magoba RNN and Samways MJ (2012) Comparative footprint of alien, agricultural and restored vegetation on surface-active arthropods. *Biological Invasions* **14**, 165–177.

McGeoch MA (2007) Insects and bioindication: theory and progress. In: *Insect Conservation Biology*. (Eds AJA Stewart, TR New and OT Lewis) pp. 144–174. CABI, Wallingford, Oxon, UK.

Mittermeier RA, Gil PR, Hoffmann M, Pilgrim J, Brooks T, Mittermeier CG, Lamoreux J and Da Fonseca GAB (2004) *Hotspots Revisited*. Cemex, Mexico City.

Pryke JS and Samways MJ (2009) Recovery of invertebrate diversity in a rehabilitated city landscape mosaic in the heart of a biodiversity hotspot. *Landscape and Urban Planning* **93**, 54–62.

Rounsevell MDA, Dawson TP and Harrison PA (2010) A conceptual framework to analyse the effects of environmental change on ecosystem sevices. *Biodiversity and Conservation* **19**, 2823–2842.

Samways MJ. (2003) Threats to the tropical island dragonfly fauna (Odonata) of Mayotte, Comoro Archipelago. *Biodiversity and Conservation* **12**, 1785–1792.

Samways MJ (2007a) Implementing ecological networks for conserving insect and other biodiversity. In: *Insect Conservation Biology*. (Eds AJSA Stewart, TR New and OT Lewis) pp. 127–143. CABI, Wallingford, Oxon, UK.

Samways MJ (2007b) Insect conservation: a synthetic management approach. *Annual Review of Entomology* **52**, 465–487.

Samways MJ, Bazelet CS and Pryke JS (2010a) Provision of ecosystem services by large-scale corridors and ecological networks. *Biodiversity and Conservation* **19**, 2949–2962.

Samways MJ, Hitchins PM, Bourquin O and Henwood J (2010b) *Tropical Island Recovery: Cousine Island Seychelles*. Wiley-Blackwell, Oxford.

Samways MJ, McGeoch MA and New TR (2010c) *Insect Conservation: A Handbook of Approaches and Methods*. Oxford University Press, Oxford.

Samways MJ and Moore SD (1991) Influence of exotic conifer patches on grasshopper (Orthoptera) assemblages in a grassland matrix at a recreational resort, Natal, South Africa. *Biological Conservation* **57**, 205–219.

Samways MJ and Sharratt NJ (2010) Recovery of endemic dragonflies after removal of invasive alien trees. *Conservation Biology* **24**, 267–277.

Samways MJ, Sharratt NJ and Simaika JP (2011) Recovery of endemic river macroinvertebrates following river bank restoration. *Biological Invasions* **13**, 1305–1324.

Simaika JP and Samways MJ (2011) Comparative assessment of indices of freshwater habitat conditions using different invertebrate taxon sets. *Ecological Indicators* **11**, 370–378.

8 LAND USE INTENSIFICATION, SMALL LANDHOLDERS, AND BIODIVERSITY CONSERVATION: PERSPECTIVES FROM THE EASTERN HIMALAYAS

Kamaljit S. Bawa, Suman Rai, Shristi Kamal and Pashupati Chaudhary

Lesson #1. Land use intensification, as practised currently, may not be a viable option at this time for small landholders in many biodiversity-rich landscapes.

Lesson #2. An alternative to land use intensification for small landholders is the diversification of livelihoods resulting in the provision of an array of products and services that can enhance economic wellbeing as well as conservation.

Lesson #3. Diversification of livelihoods to benefit local communities and local ecosystems requires integration of knowledge, new institutions for generating such knowledge, changes in policies regarding rights and tenure and strong social institutions for decentralised governance.

Lesson #4. For environmental sustainability, local land use must be integrated into regional and national level land use planning.

Lesson #5. Substantial economic, social and ecological gains can be made by implementing the concept of sustainable landscapes and by monitoring sustainability.

Introduction

Land use intensification generally refers to the intense use of land for modern agriculture dependent upon high yielding varieties with high inputs of fertilisers and pesticides (Rudel *et al.* 2009). The effects of land use intensification on biodiversity on farms are well documented (Firbank *et al.* 2008; Scherr and McNeely 2008). Indeed, land use intensification for agricultural production also impacts wild biodiversity even after allowing for certain lands to remain under agriculture. Homogenisation of the landscape, for example removal of trees and other wild plants, increases the isolation of landscapes with more or less intact biodiversity. The use of fertilisers and pesticides on agricultural lands and plantations embedded in landscapes with a high amount of wild biodiversity has adverse effects on wild biodiversity through their impact on food webs (Firbank *et al.* 2008) and pollinator services (Potts *et al.*

2010; Garibaldi *et al.* 2011). Fertilisers and pesticides also affect biodiversity by entering streams and other water bodies thus endangering fresh water biodiversity and terrestrial organisms dependent on fresh water biotas and habitats (MEA 2005; Stoate *et al.* 2009).

Many small agricultural landscapes throughout the world occur in areas extraordinarily rich in biodiversity. In many parts of the tropics, these agricultural farms are intermixed with natural habitats. Farmers often grow traditional varieties of annual crops mixed with perennial crops including trees (Nautiyal *et al.* 2003; Sirén 2007). Such crops and trees are grown without or with a minimal use of external inputs. Income from smallholdings is often insufficient to meet household needs. Therefore, income is often supplemented by wage labour and the collection of products from surrounding ecosystems for subsistence or for sale in commercial markets. Land use intensification on these farms in the form of homogenisation of the landscape and inputs of fertilisers and pesticides can have adverse consequences for on- and off-farm biodiversity.

Land use intensification may also not be economically feasible and socially desirable as the holdings are extremely small and the overall income of the farmers is very low. Globally, there are 500 million small farmers, with a majority (87%) in Asia (World Bank 2007). In India, 74% of the farmers in 1991–1992 had land holdings of less than one hectare, and this proportion now has probably increased as farm size has been declining (Fan and Chan-Kang 2005). In many parts of the Himalaya, the general location of the case study reported here, holdings may be as small as 0.25 ha. Such small landholders do not have access to credit, and in many cases neither do they have access to the markets needed for products yielded by intensive agriculture. Socially, distant markets increase the time lag between market signals and response to such signals, making local communities vulnerable to fluctuations in prices and demand for commodities.

The reliance of small farmers on local biodiversity to sustain their livelihoods in biodiversity-rich areas offers challenges as well as opportunities for increasing rural incomes and the conservation of biodiversity. The challenges are constraints to agricultural production, and with increased human populations, increased demand for land and concomitantly deleterious effects on biodiversity. The opportunities are in making agriculture and other livelihood-based activities more ecologically sustainable, diversifying the portfolio of goods and services offered by farms and forests to increase income, subsequently decreasing pressures on biodiversity and increasing the stake of farmers in conservation of biodiversity.

In this chapter, we report the preliminary results of a case study from biodiversity-rich areas in the Darjeeling Hills in the Eastern Himalayas, where we explored the potential of (1) increase in rural incomes without land use intensification; (2) diversification of products and services offered by small farms; and (3) simultaneous enhancement of gains in livelihoods and conservation of biodiversity. We first describe the setting for the study, reiterate our general objectives, outline the approach, report preliminary results and then discuss the lessons learnt.

Case study – Darjeeling District, West Bengal, India

The Darjeeling District in the state of West Bengal, India, covers an area of 3200 km² and lies in the Eastern Himalayas, a global hotspot of biodiversity. Most of the district comprises a series of hill ranges that merge with Khangchendzonga, the third highest mountain

Figure 8.1: Small-scale farming in the Eastern Himalayas. (Photo: Pashupati Chaudhary)

in the world (8586 m). Our case study comprises 15 villages in and around three protected areas: Neora Valley National Park, Senchal Wildlife Sanctuary and Singalila National Park (see Figure 8.1). Although the protected areas cover a wide range of eco-regions from tropical semi-evergreen forests to alpine shrub meadows (WWF/ICIMOD 2001), the target villages are in the Eastern Himalayan Broadleaf Forest Eco-region.

The 15 villages have a total of 474 households. The villages are small in size, with an average of 34 households. Many villages originated as forest camps when labourers were brought in for forestry operations in the early part of the last century. Forestry operations ceased towards the end of the last century and the logging camps have been transformed into forest villages that lie inside or at the periphery of protected areas. The size of most land holdings is less than 0.25 ha. The main sources of income are agriculture and livestock supplemented by wage labour and remittances. Fodder, fuel wood and other non-timber forest products for subsistence are gathered from surrounding forests.

Our objectives were to: (1) increase livelihood options, rural incomes and the local community's stake in conservation; (2) strengthen local institutions to sustain economic and conservation goals; and (3) foster the conservation of biodiversity by examining natural resource patterns and making appropriate interventions to reduce harvesting of forest products. In addition, a larger goal is to develop a sustainable landscape and the means to achieve sustainability goals. This is a long-term project, and in this chapter we report progress to date.

We have followed a multifaceted approach to enhance rural incomes by diversifying livelihood options, creating and strengthening village-based institutions to sustain new ventures and employed participative processes to monitor resource use. In addition, we have used natural and social science methods and tools to test hypotheses concerning the effects of economic, social and ecological interventions. Using a participatory approach, we have worked on rural livelihoods to make *economic gains*, on village level institutions to achieve *social gains*, and on biological resources to make *ecological gains*.

Interventions

On the *economic* front, as an alternative to land use intensification, farmers were provided training in growing high value vegetables in polyhouses for markets in towns nearby. Several farmers also received training in managing local bees for honey production. The villagers also increased the production of handicrafts based on fibre and wood from the forests. Such products are being marketed at premium prices in retail outlets especially created for the sale of handicrafts and other green products.

Since the Darjeeling region attracts a lot of tourists, some villagers offered homestays to hikers and tourists. The Forest Department further developed this idea, and built a tourist lodge near one of the villages. However, we are uncertain if the lodge will benefit the villagers because of the centralised manner in which the state enterprises are run.

Our preliminary analysis of data indicates that interventions by the Ashoka Trust for Research in Ecology and the Environment (ATREE) have increased the average income of participating villagers by about 7% above the overall group in one year but there is potential for much greater increases. Moreover, the 7% figure is average for about 140 households. The range in increase is huge, with many households increasing their income by more than 200%.

Finally, a retail outlet for handicrafts and other products in Darjeeling, the largest town in the region, has benefited hundreds of villagers.

On the *social* front, we have created or strengthened a number of institutions to sustain economic activities. These institutions range from self-help groups created in the communities to carry out income generation activities to a retail outlet established in Darjeeling town to sell handicrafts. More than half of the 33 self-help groups are all-women groups. These groups are linked with small banks to enable them to access loans.

The *ecological* gains have been mostly in the form of reducing harvesting of forest products because of the potential of other enterprises to yield income. Bio-briquettes have been introduced in a few villages as a trial to reduce fuel wood burning. The villagers have made an effort to restore degraded land. However, we do not know if the villagers' impacts on the forests have been reduced. We are currently involved in quantifying the forest resources gathered, and are initiating a comprehensive biodiversity monitoring program with the participation of local communities.

Table 8.1 shows the ATREE intervention linkages between the goals and the outcomes.

Lessons

1. Land use intensification, as practised currently, is not a viable option at this time for small landholders in biodiversity-rich landscapes

In our case study area, and we believe that this is true for millions of small farmers in biodiversity-rich landscapes, application of fertilisers and pesticides on crops will have

Table 8.1:   Linkages of ATREE intervention with its goals and outcomes

Goals	Interventions	Outcomes
Economic	High value crop production Beekeeping and honey production Training and marketing of handicrafts Homestays Retail shop	Rural income increased Market linkages improved Markets secured through retail shops
Social	Local institutions (Self-Help Groups) Cooperative marketing Creation of linkages with banks	Local capacity enhanced and social capital increased
Ecological	Bio-briquette making and distribution Plantations in degraded lands Participatory monitoring	Fuel wood collection reduced Degraded land restored Environmental stewardship built

deleterious consequences for biodiversity in surrounding landscapes. Small farmers also lack capital and credit is not easily available for purchasing the inputs required for intensification of land use. Moreover, the access to distant markets is difficult and attempts to intensify land use also can increase economic vulnerability.

Eco-agriculture or agriculture based on traditional crops and traditional land use practices, and principles of agroforestry can not only increase ecosystem services and productivity on farms (Scherr and McNeely 2008), but also have a positive impact on wild biodiversity. Furthermore, low-energy-input agriculture will increasingly become important for climate change mitigation.

2. An alternative to land use intensification for small landholders is the diversification of livelihoods resulting in the provision of an array of products and services that can enhance economic wellbeing as well as conservation

We have shown at several sites that the income of small landholders can be increased in several ways (Bawa *et al.* 2007; Kamal *et al.* 2011). Certainly the small size of holdings places severe limitations to growing diverse and sufficient foods, but if farming is integrated with other activities such as beekeeping and homestays that depend upon local biodiversity, then not only can income be increased but also the farmers' stake in conserving biodiversity will be enhanced. Further increases in income are possible through payment for ecosystem services and through the UN's Collaborative Programme REDD plus (Reducing Emissions from Deforestation and Forest Degradation in Developing Countries), provided suitable mechanisms for the distribution of such payments as well as monitoring of biodiversity can be developed (Phelps *et al.* 2010).

3. Diversification of livelihoods to benefit local communities and local ecosystems requires integration of knowledge, new institutions for generating such knowledge, changes in policies regarding rights and tenure and strong social institutions for decentralised governance

The success of our approach to improve rural livelihoods and to meet the multiple goals of economic and social wellbeing, sustainability of rural development initiatives and conservation of biodiversity – all at the same time is a challenge. Such a challenge can be met if a number of conditions can be fulfilled.

First is the integration of different knowledge systems. There is often considerable local knowledge about the use of land, practising traditional cropping patterns and the use of local ecosystem resources. Such knowledge, when combined with modern new knowledge about socio-ecological systems and information about pressures on land use (which will be exacerbated by climate change), can result in the development of alternatives to land use intensification.

Second, as we have argued earlier, the integrated knowledge needed to address issues of development and environment may require the development of new institutions, both formal and informal. This is particularly true in developing countries, unless the existing institutions can rise to the occasion to meet contemporary challenges (Bawa *et al.* 2008).

Third, livelihoods in biodiversity-rich areas cannot be separated from the surrounding local ecosystems from which communities derive a variety of ecosystem services, and over which local communities have had formal or informal rights. Such rights have often not been recognised, settled or devolved to local communities. Unless these rights are settled and joint plans are developed to manage land and natural resources, it will be difficult to make socio-economic gains as well as ecological gains (Bawa *et al.* 2010).

Finally, strong village-level institutions will be needed to meet livelihood needs from the development of multiple goods and services from small areas of land without endangering biodiversity, and from adopting diversified livelihoods and income-generating activities. Continuing and rapid evolution towards decentralised governance will be critical for success in establishing and sustaining such institutions.

4. For environmental sustainability, local land use must be integrated into regional and national level land use planning

A logical extension of lessons 1–3 is that district or state-level land use planning must take into account the multiple, and often seemingly conflicting, needs of agricultural production, livestock rearing, rural energy, nature tourism and biodiversity conservation. Land use planning is currently sectoral, and rarely involves local stakeholders.

5. Substantial economic, social and ecological gains can be made by implementing the concept of sustainable landscapes and by monitoring sustainability

Developing landscapes that can meet future needs without compromising current needs and that can meet the United Nations Millennium Development Goals will remain a challenge unless the land's potential to fulfil multiple needs locally and at larger scales is assessed and monitored (Scherr and McNeely 2008). Currently mechanisms for such assessments and monitoring are non-existent (Bawa 2010). Local land use planning at the district level and lower administrative levels combined with participatory monitoring have considerable potential in making the concept of sustainable landscapes a reality.

Conclusion

With continuing growth in population and consumption, pressure on biodiversity-rich lands will keep rising. Intensification of land use in these areas will have deleterious consequences for biodiversity in agroecosystems as well as wild lands. Alternatives to land use intensification on such lands in the tropics could be the optimal use of land by farmers through diversification of livelihoods using both cultivated and wild lands in a manner that enhances the prospect of conservation by ecological, economic and social transformations. Such transformations will require agricultural innovations, integration of knowledge systems, participatory local land use planning and enabling institutions and policies. Our case study demonstrates that changes are feasible, and can be accelerated with benefits for both nature and society.

Acknowledgements

Our work in the Himalayas has been supported by grants from the Blue Moon Fund, Ford Foundation, MacArthur Foundation, and the Sehgal Family Foundation.

Biographies

Kamaljit S. Bawa is a Distinguished Professor of Biology at the University of Massachusetts Boston, and President of the Ashoka Trust for Research in Ecology and the Environment (ATREE) in Bangalore, India. He has edited or authored 10 books and published more than 180 papers in ecology, conservation biology and sustainability science. Kamal is also editor-in-chief of *Conservation and Society*.

Pashupati Chaudhary is an environmental scientist working for ATREE. He earned his PhD degree from the University of Massachusetts, Boston, USA. He has published many papers in scientific journals, among them *Biology Letters*, *Current Science*, and *Journal of Applied Geography*.

Suman K. Rai is working with ATREE in India. He has a Master's degree in Economics from North Bengal University in India. Suman has worked extensively across eight countries of the Hindu Kush Himalayas. His work has focused on the management of natural resources and an inclusive approach to conservation. During recent years Suman has been working with grassroots communities in north-east India.

Shristi Kamal is the Program Coordinator for the Conservation and Livelihoods Program of ATREE in the Eastern Himalaya. She has a Master's degree from TERI University. Shristi is from Assam where she has also extensively volunteered for local organisations working for the conservation of biodiversity.

References

Bawa KS (2010) Monitoring systems outdated and protectionist. *Nature* **466**, 920.
Bawa KS, Rai ND and Sodhi NS (2010) Rights, governance, and conservation of biological diversity. *Conservation Biology*, **25**(3), 639–641.

Bawa KS, Raven P and Balachander G (2008) A case for new institutions. *Science* **319**(5860), 136.

Bawa KS, Joseph G and Setty S (2007) Poverty, biodiversity and institutions in forest-agriculture ecotones in the Western Ghats and Eastern Himalaya ranges of India. *Agriculture, Ecosystems & Environment* **121**(3), 287–295.

Fan S and Chan-Kang C (2005) Is small beautiful? Farm size, productivity, and poverty in Asian agriculture. *Agricultural Economics* **32**, 135–146.

Firbank LG, Petit S, Smart S, Blain A and Fuller EJ (2008) Assessing the impacts of agricultural intensification on biodiversity: a British perspective. *Philosophical Transactions of Royal Society B* **363**(1492), 777–787.

Garibaldia LA, Aizena MA, Klein AM, Cunningham SA and Harder LD (2011) Global growth and stability of agricultural yield decrease with pollinator dependence. *Proceedings of National Academy of Sciences* **108**, 5909–5914.

Kamal S, Rai S and Bawa KS (2011) 'Enhancing conservation and livelihood security in biodiversity hotspots, mountain biodiversity and conservation: selected examples of good practices and lessons learned from the Hindu Kush–Himalayan Region'. ICIMOD, Kathmandu, Nepal.

Millennium Ecosystem Assessment (MEA) (2005) *Ecosystems and Human Well-Being: Synthesis*. Island Press, Washington.

Nautiyal S, Maikhuri RK, Rao KS, Semwal RL and Saxena KG (2003) Agro-ecosystem function around a Himalayan Biosphere Reserve. *Journal of Environmental Systems* **29**, 71–100.

Phelps J, Webb EL and Agrawal A (2010) Does REDD+ threaten to recentralize forest governance? *Science* **328**, 312–313.

Potts SG, Biesmeijer JS, Kremen C, Neumann P, Schweiger O and Kunin WE (2010) Global pollinator declines: trends, impacts and drivers. *Trends in Ecology and Evolution* **25**, 345–353.

Rudel TK, Schneider L, Uriarte M, Turner BL 2nd, DeFries R, Lawrence D, Geoghegan J, Hecht S, Ickowitz A, Lambin EF, Birkenholtz T, Baptista S and Grau R (2009) Agricultural intensification and changes in cultivated areas, 1970–2005. *Proceedings of National Academy of Sciences* **106**, 20675–20680.

Scherr SJ and McNeely JA (2008) Biodiversity conservation and agricultural sustainability: towards a new paradigm of 'eco-agriculture' landscapes. *Philosophical Transactions of Royal Society B* **363**(1491), 477–494.

Sirén AH (2007) Population growth and land use intensification in a subsistence-based indigenous community in the Amazon. *Human Ecology* **35**(6), 669–680.

Stoate C, Ili AB, Beja P, Boatman ND, Herzon I, van Doorn A, de Snoo GR, Rakosy L and Ramwell C (2009) Ecological impacts of early 21st century agricultural change in Europe – A review. *Journal of Environmental Management* **91**, 22–46.

World Bank (2007) 'World Development Report 2008: Agriculture for Development'. World Bank, Washington.

WWF, ICIMOD (2001) 'Ecoregion-based conservation in the Eastern Himalaya: identifying important areas for biodiversity conservation'. Kathmandu: WWF Nepal program.

9   RICHES TO RAGS: THE ECOLOGICAL CONSEQUENCES OF LAND USE INTENSIFICATION IN NEW ZEALAND

Raphael K. Didham, Lisa H. Denmead and Elizabeth L. Deakin

Lesson #1. The agricultural transformation of New Zealand is a recent historical phenomenon, with a colonial agricultural model impressed onto a region with a distinct biogeographic and evolutionary history, making direct comparisons difficult with land use impacts in ancient agroecosystems of the northern hemisphere.

Lesson #2. Local trajectories of land use change increasingly depend on global market forces.

Lesson #3. Definitional conflicts over what constitutes intensification versus extensification have obscured debate over the ecological impacts of land use intensification.

Lesson #4. Ecological impacts are likely to scale linearly for on-site effects and non-linearly for off-site effects with increasing land use intensification.

Lesson #5. The magnitude of land use impact on biodiversity depends on the degree of structural or environmental contrast between production land use and historical landscape cover.

Lesson #6. A robust reserve network on unproductive land is not sufficient to offset biodiversity losses from land use intensification on productive land.

Lesson #7. The spatial scales of management actions do not match the spatial scales of ecological responses to agricultural intensification.

Lesson #8. The relative gain from conservation actions depends sensitively on adjacent land use intensity.

Introduction

New Zealand is a land of environmental contradictions. Despite having a conservation reserve network that is the envy of the world, unrestricted land use intensification has greatly exacerbated conservation threats in 'the other 60%' of the country used for agricultural production (McLeod *et al.* 2008). The non-random pattern of habitat modification

(Ewers *et al.* 2006) has resulted in biodiversity loss in lowland NZ that is unparalleled almost anywhere in the world (Norton and Miller 2000). Nearly half of the administrative districts of NZ have suffered >85–90% conversion to agriculture, and the remaining natural vegetation cover is heavily fragmented (MFE 2000; Ewers *et al.* 2006). Correspondingly, even the smallest habitat remnants have extremely high conservation value in production landscapes. For example, ~20% of NZ's threatened vascular plants are confined to private farmland and, for an additional 60%, private land constitutes a significant proportion of their habitat (Norton and Miller 2000).

Within this context, conflict has been inevitable between production and biodiversity values in modified landscapes. On the one hand, there is increasing recognition that a protected areas network built on the acquisition of high-elevation land of low commercial value can never be fully representative of NZ biodiversity. Effective conservation management must incorporate biodiversity on production lands (Norton and Miller 2000). On the other hand, the expansion of agriculture unquestionably built the modern NZ economy (accounting for >65% of exports and 17% of GDP, Rowarth *et al.* 2006). At one time, NZ agriculture also had a reputation for the best environmental performance in the world (Rowarth *et al.* 2006). However, increasing global demand for production, combined with limited available land to expand agriculture, forced an unprecedented period of rapid and unsustainable intensification of land use (PCE 2004; Rowarth *et al.* 2006; Rowarth 2008; Lee *et al.* 2008; Moller *et al.* 2008a). From 1960–2000, total production increased 200–300% for major commodities with only 6–7% increase in area farmed (MacLeod and Moller 2006; Rowarth *et al.* 2006). This trend is accelerating, and is fuelling concern among scientists, the public and politicians over how best to trade-off sustainable commodity production with effective biodiversity conservation (Lee *et al.* 2008; Moller *et al.* 2008a,b; Rowarth 2008).

Here, we present a brief précis of the socio-economic history of agricultural transformation in NZ, and its continuing influence on the natural environment. We discuss the global market forces that increasingly drive local trajectories of land use intensification in NZ, and the contentious debate over discrimination of the relative effects of land use expansion versus land use intensification, per se. Finally, we focus on emerging ecological lessons to be learned from rapid agricultural intensification in NZ and the implications of these for governance and conservation management in production landscapes.

The socio-economic context – from extractive pastoralism to productivist farming to neoliberalisation and beyond

1. The agricultural transformation of New Zealand is a recent historical phenomenon, with a colonial agricultural model impressed onto a region with a distinct biogeographic and evolutionary history, making direct comparisons difficult with land use impacts in ancient agroecosystems of the northern hemisphere

New Zealand has had a very short, but intense, agricultural history compared with many regions of the northern hemisphere, and this socio-economic and historical context has an important bearing on interpretation of ecological responses to land use intensification.

The early extractive pastoralism phase of agricultural development in NZ began in the 1840s with burning and grazing of unimproved grasslands, but only lasted a few decades before transition to pastoral farming, stimulated by international markets opening up to NZ commodities in the 1880s (MacLeod and Moller 2006). The expansion of pastoral farming caused high rates of native forest clearance, accelerated by a first wave of intensification of resource inputs from the 1920s, creating increasingly Europeanised pastoral monocultures. Strong local regulatory governance and removal of trade barriers to the UK gave privileged access to effectively unlimited markets and decoupled supply–demand dynamics for NZ farmers. New Zealand agriculture went into full-scale productivist intensification from the 1950s–1980s, which became known as the 'long boom' (Meadows et al. 2008; Haggerty et al. 2009). The post-boom agricultural crises of the 1980s crippled NZ agriculture, leading the government to experiment with various ways of restructuring the political economy. New Zealand eventually undertook the most extreme rollback of regulatory governance and agricultural subsidisation of any country in the world (by 2001 subsidies to agriculture were equivalent to <1% of agricultural output, compared with ~31% for other developed countries, Rowarth et al. 2006), and instead rolled-out comprehensive neo-liberal market reform. The consequent 'rural downturn' drove a desperate period of agricultural diversification in the 1980s–1990s, leading to decades of instability in land use and associated environmental impacts.

The ecological implications of such a recent, tumultuous agricultural transformation are threefold. First, transposition of European species and farming practices to a new biogeographic region presented novel ecological and evolutionary pressures for native species. For example, NZ forest understorey communities were not adapted to the trampling and browsing effects of hard-hoofed terrestrial mammals (Lee et al. 2008). Second, the low-intensity pastoralism phase of agricultural transformation lasted mere decades in NZ, compared to centuries in ancient European agroecosystems. This ensured rapid attrition of native biodiversity on production land with little possibility for species to respond to new selective forces. Third, given the short time frame, remnant ecosystems are inevitably paying out both a significant extinction debt (reflecting the number of species yet to become locally extinct from gradual attrition of unviable populations) and a significant invasion credit (reflecting the number of species yet to invade a remnant patch from incipient populations in the agricultural matrix) (Didham 2010). Together these unique aspects of NZ agroecosystems mean that species are likely to be more sensitive to recent land use intensification than species in ancient agroecosystems (see also Kleijn et al. 2009), and the full effect of intensification on species in adjacent semi-natural habitats is yet to be felt.

2. Local trajectories of land use change increasingly depend on global market forces

New Zealand's mixed legacy of neo-liberal reform layered onto an historical ethos of productivist agriculture has simultaneously enabled two contradictory trajectories in NZ agriculture (Haggerty et al. 2009). On the one hand, some sectors are taking the high road of good agricultural practice via highly audited environmental standards driven by consumer ethics in elite consumer markets. For example, the kiwifruit industry has effectively reversed the environmentally damaging productivism that defined agricultural transformation in the 20th century (Haggerty *et al.* 2009). Meanwhile, other sectors have spiralled further into an intensifying trajectory, driving animals and environment toward maximum production intensity because access to elite consumer markets is blocked. For example, the dairy industry responded to trade barriers imposed by wealthy markets in Europe and North America by strategically targeting the sale of low-cost, high-volume commodities to low to middle income countries of Asia and Central and South America (Jay 2005; Baskaran *et al.* 2009). Consequently, dairy farmers are forced to keep production costs low and outputs high, with little incentive to invest in environmental protection (PCE 2004; Jay 2005). The dramatic increase in land use conversion to dairy farming now means NZ supplies one-third of the global dairy trade, generating 25% of national export revenues and contributing >10% to GDP (Baskaran *et al.* 2009). This has come at the expense of exponential increases in water, energy, fertiliser, pesticide and imported feedstock use in dairy production, making dairy farming the dominant driver of land use intensification and environmental impacts in NZ today (PCE 2004; MacLeod and Moller 2006; Baskaran *et al.* 2009). Ultimately, this stems from neo-liberal adoption of private-sector governance over food chains (Haggerty *et al.* 2009), which has gone awry because global market forces are decoupled from local ecological feedbacks on production or market access (Campbell 2009). Understanding that decoupling is critical to understanding the drivers of environmental change, and reasserting that coupling will be critical to sustainable agricultural development and biodiversity conservation.

Land use intensification: defining the problem

3. Definitional conflicts over what constitutes intensification versus extensification have obscured debate over the ecological impacts of land use intensification

The distinction between expansion of agriculture across a larger total land area versus intensification of agriculture per unit of land area might seem like a relatively straightforward concept, but an absolute definition of intensification has eluded both conceptual and empirical efforts to find common ground. Such definitional conflicts have led to contentious debate in NZ about the full extent of agricultural impacts, and the relative role that intensification, per se, has played in these effects (Lee *et al.* 2008; Moller *et al.* 2008b).

Krebs *et al.* (1999, p. 611) considered that 'intensification is about making as great a proportion of primary production as possible available for human consumption'. This echoes one standard definition of agricultural intensification as increasing production output per unit area per unit time (Turner and Doolittle 1978). However, ecologists have applied the term more loosely to refer to the state of agricultural mechanisation or amount of farm inputs rather than outputs (Kleijn *et al.* 2009). The rationale is that inputs are 'more likely to be related to biodiversity than yield per se' (Kleijn *et al.* 2009, p. 904). By contrast, the Food and Agriculture Organization (FAO) has a distinctly different socio-economic definition of intensification that is in direct conflict with the ecological perspective:

' ... an increase in agricultural production per unit of inputs (which may be labour, land, time, fertiliser, seed, feed or cash)' (FAO 2004, p. 5). Under this definition, not just increasing land area but also farm inputs are synonymous with expansion of agriculture (FAO 2004). Consequently, intensification is more a statement of economic efficiency of production, and confounds any ecological impacts stemming from greater inputs or outputs per unit land area (Moller *et al.* 2008a).

These definitional problems raise at least two important ecological considerations. First, a spatially explicit definition of intensification is essential to the investigation of land use impacts because land area for organisms is not an interchangeable concept with farm inputs, as it might be in a purely economic model. Organisms vary in their distributions, and perceive and respond to spatial variation in environmental gradients across landscapes. Second, while the ecological impacts of increasing resource inputs are likely to be greater than effects stemming from maximisation of production outputs in many situations (Kleijn *et al.* 2009), this may vary between on-site and off-site ecological impacts. Many detrimental off-site impacts of intensification might actually stem from inefficiencies in the usage or targeting of inputs towards production, not from the total quantity of inputs per se. For example, large single-pulse applications of nitrogen fertiliser can result in greater loss of nitrogen to adjacent systems due to volatilisation or run-off prior to local uptake by the crop.

Ecological impacts across space and time

4. Ecological impacts are likely to scale monotonically for on-site effects and non-linearly for off-site effects with increasing land use intensification

An important shortcoming of recent studies is that the contrasting ecological mechanisms underpinning on-site versus off-site effects of intensification have not been effectively synthesised into a generalised model of the ecological dynamics of spatially coupled natural and production systems. In NZ, there is a significant amount of agronomic information about biodiversity on farmland, and even about off-site impacts on water quality and stream biodiversity in spatially coupled aquatic systems, but almost nothing is known about off-site impacts in spatially coupled terrestrial systems (Moller *et al.* 2008a). This limits a more complete understanding of the ecological impacts of intensification, and the management initiatives necessary for sustainable agricultural development. What is surprising is this is not a uniquely NZ failing, but is symptomatic of the state of knowledge of global agricultural impacts in general. Despite ample evidence that biodiversity and ecosystem services decline on farms, there had not even been any critical assessment of the actual shape of the relationship between land use intensity and biodiversity loss on farmland until recently. Across Europe, Kleijn *et al.* (2009) found a negative exponential relationship between land use intensity (farmer nitrogen inputs) and native biodiversity (on-site plant species richness). Arguably, it is a sad indictment on the state of the field that such a relationship has only just now been derived (for one region).

No comparable empirical relationship exists for NZ, but the expectation might be an equivalent monotonic relationship between intensification and on-site biodiversity when intensification is directly defined and measured in terms of the mechanism underlying ecological impact. The corresponding off-site impacts of land use intensification on biodiversity in adjacent semi-natural habitats embedded within production landscapes are even more poorly understood. However, the complexity of the relationships between farmer inputs and spillover (integrating inputs with the efficiency of farmer outputs per unit input), and between spillover and ecological impact, suggests a greater likelihood that off-site impacts will scale in a more unpredictable, perhaps non-linear, manner with increasing intensification, although this relationship is only just now being investigated in NZ (Didham, Barker, Deakin, Denmead, Schipper and Tylianakis, unpublished).

5. The magnitude of land use impact on biodiversity depends on the degree of structural or environmental contrast between production land use and historical landscape cover

It is not widely appreciated that the ecological impacts of intensification not only vary in their relative magnitude but can even show a qualitative reversal in their direction of effect depending on the type of production intensification (e.g. pastoral grazing versus plantation forestry) and the historical landscape context (e.g. intensification within grassland versus forest biomes). For example, Pawson *et al.* (2008) compared the effect of matrix land use type on beetle communities in the North Island and found that high-intensity plantations actually had higher native beetle biodiversity and lower exotic beetle invasion than low-intensity pasture, and had a beetle community that was more similar to native forest beetle composition than any other production land use. Consequently, in this landscape, ecological impact can be inferred to scale inversely with intensity of land use. In landscapes where exotic plantation forests are structurally and environmentally similar to the historical native vegetation cover, they may provide important alternative habitat for native species in spite of the higher intensity of production inputs and disturbance (Pawson *et al.* 2008). These results contrast with another study by Pawson *et al.* (2010) which experimentally tested the effects of increasing densities of exotic pine forest stands on invertebrate communities in the semi-natural grasslands of the South Island. Here, exotic forest land use stands in striking contrast with the historical biome, and the resulting land use impact on grassland invertebrate communities scales positively with increasing intensification.

Challenges in conservation management

6. A robust reserve network on unproductive land is not sufficient to offset biodiversity losses from land use intensification on productive land

The common 'land-sparing' rationale for accepting a trade-off between high agricultural productivity and low native biodiversity on production lands is that separate nature reserves can be set aside to conserve species on agriculturally marginal land (Rowarth 2008). New Zealand should be a paradigmatic example of land sparing because it has a large protected areas network, strict regulatory controls and genuine mitigation rather than displacement of threatening processes. However, the weakness of the reserve network is the lack of representativeness of protected areas across eco-regions, leaving the higher biodiversity of lowland ecosystems almost entirely unprotected (Norton and Miller 2000; Lee *et al.* 2008). Land managers now recognise that formal reserves alone are not sufficient to stem biodiversity loss (Dodd *et al.* 2011), and recent initiatives explicitly promote biodiversity conservation on private farmland (Norton and Miller 2000). Habitat remnants on farms are becoming a key focus in the management of biodiversity, and in the development of guidelines for regional biodiversity protection. For example, the Queen Elizabeth II National Trust allows landowners to voluntarily protect parts of their property in perpetuity, with over 3500 registered QEII covenants providing active management to >100 000 ha of privately owned remnants (Anon. 2008). While these figures indicate a growing desire by landowners to protect biodiversity, substantial management intervention will still be required to mitigate agricultural impacts, particularly where off-site spillover and ecological impacts scale with increasing land use intensification.

7. The spatial scales of management actions do not match the spatial scales of ecological responses to agricultural intensification

Meadows *et al.* (2008) and Moller *et al.* (2008b) identify the whole-farm enterprise as the key social-ecological scale and site of action for change in agricultural practices. While this is undoubtedly true in NZ where self-governance over private property is ingrained in the national character, the reality is that global market forces increasingly determine local farmer decisions, and those local actions have adverse landscape-scale outcomes (for other farmers and for conservation). These cross-scale cause and effect relationships are recognised, but no serious attempt has been made to address them, and there remains a significant mismatch between the spatial scales of ecological responses to land use intensification and the spatial scales of management actions. Whether market forces alone can address these mismatches is heavily debated (Lee *et al.* 2008; Meadows *et al.* 2008; Moller *et al.* 2008a,b), but seems unlikely, and government and local authorities seem reluctant to deploy available regulatory tools. For instance, under the Resource Management Act (RMA) and Long Term Council Community Plans, local authorities have the capacity to set limits on land use types and farming intensity, like those operating in the European Union, but they are almost never implemented. Some Regional Councils have also considered making the whole-farm operation a formally consented activity under the RMA to prevent off-site impacts from high intensity land use (Meadows *et al.* 2008), but this has never been implemented either. A key challenge for sustainable agriculture will be to match effective landscape-level management to the landscape-scale ecological impacts of land use intensification.

8. The relative gain from conservation actions depends sensitively on adjacent land use intensity

Just as the costs of conservation actions vary with spatial context (e.g. the costs of land acquisition are lower for marginal agricultural land), so too do the benefits. For example, although fencing to exclude livestock from habitat remnants might cost a similar amount per metre of fence regardless of land use intensity, the relative conservation gain could differ dramatically depending on the rarity of habitat type or level of habitat degradation under high-intensity land uses. In general, the degree of context-dependence in the cost to benefit ratio of conservation actions will depend on the scaling relationship between land use intensity and ecological impacts. If biodiversity loss scaled linearly with intensification, a given unit of conservation management effort would be just as effective at all levels of intensification, whereas if biodiversity loss scaled exponentially, or as a threshold function of intensification, then management actions might be less effective at high land use intensity. For instance, in the Kleijn *et al.* (2009) scenario for Europe, where there was an exponential decline in native plant diversity on farms with increasing fertiliser inputs, conservation initiatives might be most effective, dollar-for-dollar, if they are preferentially implemented on extensively managed farms that still support high levels of biodiversity. Paradoxically, this is only going to widen the gulf that already exists between land allocated to intensive agriculture versus land set aside for nature conservation. Furthermore, recent studies suggest that there is only relatively weak evidence for net biodiversity benefits associated with this 'land sparing' rationale to land use intensification (Ewers *et al.* 2009), and it runs counter to the growing recognition of the importance of maintaining biodiversity within production landscapes (Norton and Miller 2000).

Conclusions

There is no question that agriculture has transformed, and continues to transform, the NZ landscape. For ecologists, determining the relationship between land use intensification and biodiversity loss, both on the farm and in adjacent semi-natural ecosystems, represents the cornerstone of sustainable agricultural development in the future. For land managers, a first step must be to address the coupling of global market forces and local environmental outcomes. At present, important sectors of NZ agriculture are in a negative spiral of land use intensification and environmental degradation stemming from the global corporate model of privatisation of profits and socialisation of costs. We would argue that the NZ public are simply subsidising the profits that companies make from exporting the products of high-intensity land use, by internalising the costs of environmental degradation within the country. One way or another, these environmental costs will ultimately be paid through loss of biodiversity, increased costs of environmental restoration or the tarnishing of the tourism industry's 'clean, green' PR slogan. A second important step will be to develop landscape-scale management strategies that integrate the need to extract resources, control pest species and sustain livelihoods on the farm, with the need to maintain populations of native and beneficial introduced species in spatially coupled natural and production landscapes (Moller *et al.* 2008a). These strategies must recognise the off-site effects of land use intensification. This represents a crucial missing piece of the puzzle in terms of understanding the trade-offs to be made between agri-environment schemes that aim to reduce land use intensification, and land-sparing arguments that promote the localised increase in land use intensification in some areas to offset preservation of 'wild nature' in other areas.

Acknowledgements

We thank Saul Cunningham and David Lindenmayer for constructive comments on earlier versions of this article. This work was supported by the Royal Society of New

Zealand Marsden grant UOC0803 'Riches to rags: does elevated productivity drive ecosystem decay in adjacent natural habitats?' We acknowledge our collaborators in the 'riches to rags' project, Gary Barker, Scott Bartlam, Louis Schipper and Jason Tylianakis.

Biographies

Raphael K. Didham is Professor of Ecology in the School of Animal Biology at the University of Western Australia, and holds a joint research position at CSIRO Ecosystem Sciences, Perth, Australia. Raphael received his PhD from Imperial College London in 1997 and completed a postdoctoral fellowship at the University of Delaware, USA, and a teaching fellowship with Boston University and the School for Field Studies in Canada, before holding a faculty position at the University of Canterbury, in NZ, for the past 10 years. The goal of Raphael's research is to quantify the synergistic effects of multiple drivers of global change on biodiversity and ecological resilience of remnant natural ecosystems within production landscapes, with a particular focus on conserving invertebrate biodiversity and maintaining natural ecosystem services.

Lisa H. Denmead is a MSc candidate at The University of Western Australia, Perth, Australia, and received a BSc in Biology from the University of Canterbury, New Zealand. Lisa's interests are focused on investigating invertebrate biodiversity, community dynamics and ecosystem processes in remnant natural ecosystems. Her research is testing whether the relative gain from conservation actions depends on adjacent land use intensity.

Elizabeth L. Deakin is a PhD candidate at the University of Canterbury, Christchurch, New Zealand. Liz's PhD research aims to quantify the ecological impacts of agricultural intensification on spatially coupled native forest fragments, with a particular focus on plant–insect interactions. Liz holds a BSc (Hons) degree in Environmental Management and a MSc degree in Conservation from University College London. She has previously worked on numerous ecological research projects in Indonesia, Cambodia, Central America and the Mediterranean, as well as for two UK-based environmental consultancies. Liz's research interests focus on the effects of land use change on biodiversity and ecosystem functioning, with the aim of developing management tools to conserve natural habitats under anthropogenic pressure.

References

Anon. (2008) QEII Trust: helping you protect the special character of your land. *Open Space: Magazine of the Queen Elizabeth II National Trust* **74**, 31.

Baskaran R, Cullen R and Colombo S (2009) Estimating values of environmental impacts of dairy farming in New Zealand. *New Zealand Journal of Agricultural Research* **52**(4), 377–389.

Campbell H (2009) Breaking new ground in food regime theory: corporate environmentalism, ecological feedbacks and the 'food from somewhere' regime? *Agriculture and Human Values* **26**(4), 309–319.

Didham RK (2010) Ecological consequences of habitat fragmentation. In *Encyclopedia of Life Sciences*. John Wiley & Sons, Ltd, Chichester.

Dodd M, Barker G, Burns B, Didham RK, Innes J, King C, Smale M and Watts C (2011) Resilience of New Zealand indigenous forest fragments to impacts of livestock and pest mammals. *New Zealand Journal of Ecology* **35**(1), 83–95.

Ewers RM, Kliskey AD, Walker S, Rutledge D, Harding JS and Didham RK (2006) Past and future trajectories of forest loss in New Zealand. *Biological Conservation* **133**(3), 312–325.

Ewers RM, Scharlemann JPW, Balmford A and Green RE (2009) Do increases in agricultural yield spare land for nature? *Global Change Biology* **15**(7), 1716–1726.

FAO (2004) 'The ethics of sustainable agricultural intensification.' FAO Ethics Series 3. Food and Agriculture Organization of the United Nations, Rome.

Haggerty J, Campbell H and Morris C (2009) Keeping the stress off the sheep? Agricultural intensification, neoliberalism, and 'good' farming in New Zealand. *Geoforum* **40**(5), 767–777.

Jay M (2005) Remnants of the Waikato: native forest survival in a production landscape. *New Zealand Geographer* **61**, 14–28.

Kleijn D, Kohler F, Baldi A, Batáry P, Concepcion ED, Clough Y, Diaz M, Gabriel D, Holzschuh A, Knop E, Kovacs A, Marshall EJP, Tscharntke T and Verhulst J (2009) On the relationship between farmland biodiversity and land-use intensity in Europe. *Proceedings of the Royal Society B Biological Sciences* **276**(1658), 903–909.

Krebs JR, Wilson JD, Bradbury RB and Siriwardena GM (1999) The second silent spring? *Nature* **400**(6745), 611–612.

Lee WG, Meurk CD and Clarkson BD (2008) Agricultural intensification: whither indigenous biodiversity? *New Zealand Journal of Agricultural Research* **51**(4), 457–460.

MacLeod CJ, Blackwell G, Moller H, Innes J and Powlesland R (2008) The forgotten 60%: bird ecology and management in New Zealand's agricultural landscape. *New Zealand Journal of Ecology* **32**(2), 240–255.

MacLeod CJ and Moller H (2006) Intensification and diversification of New Zealand agriculture since 1960: an evaluation of current indicators of land use change. *Agriculture Ecosystems & Environment* **115**(1–4), 201–218.

Meadows S, Gradwohl M, Moller H, Rosin C, MacLeod CJ, Weller F, Blackwell G and Perley C (2008) Pathways for integration of biodiversity conservation into New Zealand's agricultural production. *New Zealand Journal of Agricultural Research* **51**(4), 467–471.

MFE (2000) 'Final report of the Ministerial Advisory Committee on biodiversity and private land'. Ministry for the Environment: Wellington, New Zealand.

Moller H, MacLeod CJ, Haggerty J, Rosin C, Blackwell G, Perley C, Meadows S, Weller F and Gradwohl M (2008a) Intensification of New Zealand agriculture: implications for biodiversity. *New Zealand Journal of Agricultural Research* **51**(3), 253–263.

Moller H, Blackwell G, Weller F, MacLeod CJ, Rosin C, Gradwohl M, Meadows S and Perley C (2008b) Social-ecological scales and sites of action: keys to conserving biodiversity while intensifying New Zealand's agriculture? *New Zealand Journal of Agricultural Research* **51**(4), 461–465.

Norton DA and Miller CJ (2000) Some issues and options for the conservation of native biodiversity in rural New Zealand. *Ecological Management and Restoration* **1**, 26–34.

Pawson SM, Brockerhoff EG, Meenken ED and Didham RK (2008) Non-native plantation forests as alternative habitat for native forest beetles in a heavily modified landscape. *Biodiversity and Conservation* **17**(5), 1127–1148.

Pawson SM, McCarthy JK, Ledgard NJ and Didham RK (2010) Density-dependent impacts of exotic conifer invasion on grassland invertebrate assemblages. *Journal of Applied Ecology* **47**(5), 1053–1062.

PCE (2004) 'Growing for good: intensive farming, sustainability and New Zealand's environment'. Parliamentary Commissioner for the Environment, Wellington, New Zealand.

Rowarth JS (2008) Agricultural intensification protects global biodiversity. *New Zealand Journal of Agricultural Research* **51**(4), 451–455.

Rowarth JS, Caradus JR and Goldson SL (2006) Agriculture: a growing concern. *Clean Air and Environmental Quality* **40**, 33–39.

Turner BL and Doolittle WE (1978) The concept and measure of agricultural intensity. *Professional Geographer* **30**(3), 297–301.

10 LAND USE INTENSIFICATION AND THE STATUS OF FOCAL SPECIES IN MANAGED FOREST LANDSCAPES OF NEW BRUNSWICK, CANADA

Marc-André Villard

Lesson #1. Intensive conifer plantations increase the predation risk on bird nests in adjacent habitat patches.

Lesson #2. Conifer plantations may alter functional connectivity among populations of some deciduous forest songbirds.

Lesson #3. Partial harvesting in broadleaf deciduous stands causes habitat degradation, but creates no ecological traps, for focal songbird species.

Lesson #4. Partial harvesting in broadleaf deciduous stands reduces the abundance and fertility of some epiphytic macrolichens.

Lesson #5. Certain songbirds exhibit species-specific threshold responses to landscape composition that are consistent through time.

Lesson #6. Threshold responses of mature forest songbirds to landscape composition vary geographically, but the probability of occurrence of species increases within larger habitat blocks.

Lesson #7. In forest bird assemblages of regions characterised by fine-scale natural disturbances, species sensitive to contemporary forest management are more common than in regions with higher-intensity natural disturbance regimes.

Introduction

Forest landscapes managed for timber are generally thought to maintain a higher level of ecological integrity than those whose matrix is dominated by agriculture, probably because in farmland, the matrix provides more resources to generalist predators than that of managed forest landscapes (Andrén 1992). Yet, intensive plantation silviculture has the potential to alter trophic relationships or even provoke trophic cascades through artificially high resource pulses (cycles of high and low food abundance). The province of New

Brunswick features some of the most intensively managed forest landscapes in Canada (MacLean *et al.* 2002). Conifer plantation establishment was undertaken in the late 1950s and the provincial government is currently considering the option of devoting up to 28% of public forest lands to spruce plantation. This expansion would be achieved by allowing spruces to be planted in clear-cut stands whose conifer regeneration (>60% stocking) used to be considered sufficient from a forestry perspective (Steve Gordon and Dan Beaudette, NB-DNR, *pers. comm.*).

Intensive conifer plantation silviculture may not only affect local biodiversity. It may also influence species inhabiting adjoining habitat patches by generating resource pulses (e.g. high and low cone crops) that may in turn promote an influx of seed-eating rodents during years of low cone production. This may influence songbird reproduction or populations of generalist predators. Trophic cascades associated with mast-seeding events in oaks are well documented (e.g. Schmidt *et al.* 2008) but surprisingly few studies to date have investigated the potential effect of tree planting on landscape-level trophic cascades. Another important issue in north-eastern North American broadleaf deciduous forests pertains to the effects of partial harvest treatments. Although partial harvest systems could potentially be designed to meet both commercial and conservation values simultaneously, they also may inadvertently degrade habitat or even create ecological traps, i.e. habitat that is attractive for certain organisms, in spite of its lower quality (e.g. Robertson and Hutto 2007). In this chapter, I will draw lessons from our research in Canada on the effects of partial harvesting of broadleaf deciduous forests and intensive conifer plantation in the matrix on focal species of songbirds and epiphytic lichens.

Lessons

1. Intensive conifer plantations increase the predation risk on bird nests in adjacent habitat patches

We hypothesised that predation risk on songbird nests would decrease with distance from the nearest cone-producing spruce plantations, owing to the effect of cone crops on populations of rodents, particularly the Red Squirrel (*Tamiasciurus hudsonicus*), a frequent nest predator. We tested this hypothesis using artificial ground and shrub nests, as well as natural nests of Brown Creeper (*Certhia americana*). There was no evidence for an edge effect on artificial nest predation (Carignan and Villard 2002), nor was there an effect of proximity to the nearest plantation edge on creeper nest predation (Poulin and Villard 2011). However, proximity to hard edges in general and the proportion of cone-producing plantations within a 2-km radius around a nest were the most important predictors of nest predation risk (Poulin and Villard 2011). The Red Squirrel is a confirmed predator on Brown Creeper nests (Poulin, D'Astous and Villard, unpublished data) and its survival rate is strongly influenced by cone crops (Halvorson and Engeman 1983). Hence, it seems plausible that increased cone production from extensive spruce plantations would promote secondary pulses in seed-eating rodents, which in turn would provoke large fluctuations in the reproductive success of certain songbird species, as observed in the Brown Creeper by Poulin *et al.* (2010).

2. Conifer plantations may alter functional connectivity among populations of some deciduous forest songbirds

Many bird species, whether resident or migratory, have been shown to be reluctant to cross gaps of inhospitable habitat (Bélisle and Desrochers 2002; Gillies and St. Clair 2008; Robertson and Radford 2009). To assess the permeability of our managed forest landscape to songbird movements, we have performed experimental translocations on Ovenbirds (*Seiurus aurocapilla*). Translocations were conducted in two contrasting landscapes: one dominated by relatively young (ca. 25-year-old) spruce plantations and the other by partially harvested deciduous stands. Ovenbirds are typically associated with closed canopy, broadleaf deciduous stands. Preliminary results indicate that return rate of translocated birds was significantly higher in a matrix of partially harvested deciduous stands than in one dominated by spruce plantations. We repeated this experiment in 2011 and found the same effect.

3. Partial harvesting in broadleaf deciduous stands causes habitat degradation, but creates no ecological traps, for focal songbird species

Relative to uncut plots, all forms of partial harvesting (single-tree selection, patch cutting, and strip cutting) used to manage broadleaf deciduous forests in New Brunswick and elsewhere in north-eastern North America result in density reductions for focal species such as the Ovenbird (Pérot and Villard 2009) and the Brown Creeper (Poulin *et al.* 2010). These two focal species, whose populations we have now monitored for several years, were independently shown to be the two songbird species most sensitive to partial harvesting in North America (Vanderwel *et al.* 2007). In contrast with other study systems (e.g. forest thinning and Olive-sided Flycatcher, *Contopus cooperi*; Robertson and Hutto 2007), we found no evidence for maladaptive habitat selection or ecological traps (sensu Pärt *et al.* 2007). Neither focal songbird species appeared to exhibit a preference for partial cuts relative to control plots. In the Ovenbird, density was consistently lower in partial cuts than in controls (Haché and Villard 2010) and appeared to be adjusted to the intensity of harvest treatments (Pérot and Villard 2009). Upon completion of spring migration, males occupied control plots up to 5 days earlier, suggesting a preference for untreated forest stands. Although partial harvesting allows combining, to some extent, commercial and conservation objectives, its cumulative effects at the landscape scale should be carefully considered because a high proportion of broadleaf deciduous forest is managed under partial harvest systems.

4. Partial harvesting in broadleaf deciduous stands reduces the abundance and fertility of some epiphytic macrolichens

Epiphytic lichens have been recognised as indicators of forest integrity in many regions of the world (e.g. Pykälä 2004). We examined the response of two large foliose lichen species (*Lobaria pulmonaria* and *L. quercizans*) growing on trees to a selection harvest treatment removing 30–40% of the basal area. Stands had been harvested through single-tree selection 5–9 years prior to the study. Mature stands that had been left undisturbed for at least 35 years acted as controls. Although both species had persisted after harvesting, their abundance was 4–5 times higher in control stands. Most importantly, fertile *L. quercizans* and *L. pulmonaria* were respectively 5 and 26 times more frequent in control stands (Edman *et al.* 2008).

5. Mature forest songbirds generally exhibit species-specific threshold responses to landscape composition that appear to be highly consistent through time

Matthew Betts and I have investigated landscape-level thresholds in forest songbird occurrence in two managed forest landscapes of New Brunswick to determine (1) whether some species actually exhibited threshold responses; and if so, (2) whether these thresholds were broadly consistent through time and space. Results indicate that most of the mature forest bird species examined exhibit threshold responses to the amount of suitable habitat at the landscape scale (Betts *et al.* 2007), and that landscape thresholds are remarkably consistent between years in a given study area (Betts and Villard 2009 – Figure 10.1). Therefore, it appears that even highly mobile species like forest songbirds benefit from the presence of blocks of suitable habitat. When habitat amount falls below species-specific thresholds, the probability of finding these species falls at a fast rate. This means that land use intensification must be done strategically so that relatively unbroken blocks of suitable habitat are maintained in the landscape.

6. Threshold responses of mature forest songbirds are geographically variable, but species occurrence increases within larger habitat blocks

When comparing thresholds for a generic landscape variable (% mature forest) between two study areas located ca. 250 km apart in New Brunswick, we found no evidence for geographically consistent responses, except for one species (Blackburnian Warbler, *Dendroica fusca*) (Betts and Villard 2009). Species-specific values for habitat thresholds at the landscape scale were variable (13–45%), but they indicated that probability of occupancy increased significantly as the area of suitable habitat increased within a radius of 500 m to 2 km (Betts *et al.* 2007), even when controlling for spatial autocorrelation in species distribution. Therefore, thresholds do not seem to provide conservation targets that can be extrapolated across distant study areas, but our results indicate that the focal species' probability of occurrence increased consistently with the area of mature forest, irrespective of location.

7. In forest bird assemblages of regions characterised by fine-scale natural disturbances, species sensitive to contemporary forest management are more common than in regions with higher-intensity natural disturbance regimes

Pierre Drapeau, myself and two colleagues compared the response of forest songbirds to harvesting intensity across four forest regions of Canada with contrasting natural disturbance regimes. The extremes were the Alberta boreal mixedwood forest, where forest fires are frequent and spatially extensive, and the northern hardwood forest of New Brunswick, which is characterised by fine-scale 'gap-replacing disturbances', such as tree senescence, disease, and windthrow killing one tree or a small cluster at a time. We predicted that

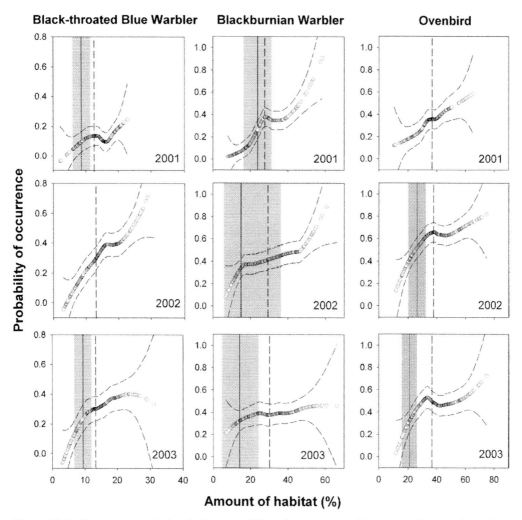

Figure 10.1: Year-to-year variation in the probability of occurrence of three songbird species in the same study area as a function of amount of suitable habitat within a 1-km (Blackburnian Warbler, *Dendroica fusca*) or 2-km radius around sampling points. Surveys were conducted at different sampling points each year. Dots represent values estimated using logistic regression models, along with confidence intervals. Solid lines correspond to thresholds obtained from segmented logistic regression, whereas dashed lines represent thresholds obtained using ROC-analysis (Betts and Villard 2009).

species whose occurrence is positively correlated with contemporary cover of mature forest at the landscape scale (*sensitive* species) would be less common (as shown by frequency rank) in forest bird assemblages of regions with historically high natural disturbance regimes, and would occupy a higher rank in regions with low disturbance regimes. Our results were consistent with this prediction. In addition, the weight of sensitive species (proportion of all individuals detected) increased faster with mature forest cover in New Brunswick than in Alberta.

Conclusions

New Brunswick has a 300-year history of commercial forestry. In Canada, it was the first province to practice plantation silviculture over significant areas, starting in the late 1950s. Large-scale partial harvesting in the hardwoods started in the 1980s. Since then, forestry intensity has increased dramatically, especially in hardwood stands (Erdle 2008). Virtually all forest land is under some form of management, with strictly protected areas representing less than 5% of the land base. As a province, it can be seen as a microcosm of Canada, with a low human population and infrastructures concentrated in a few cities in the south, and the bulk of the land under some form of resource management in the north.

New Brunswick's forest management strategy is inspired by the triad approach, whereby the public land base is allocated to intensive conifer silviculture, extensive forest management (e.g. partial harvest systems) and temporarily or permanently protected areas. Temporary protected areas include deer wintering yards, riparian buffer strips, and patches meeting the definition of 'old forest' based on their stand structure (NB-DNR 2005). This management approach is essentially 'aspatial', except for occasional guidelines about patch size for old forest.

In this context, the findings on landscape thresholds in forest bird occurrence point towards a revision of the forest management strategy to protect or restore large blocks of mature/old forest. Attention should also be paid to ensure that these forest blocks are not located near extensive conifer plantations (Poulin and Villard 2011). Furthermore, old forest definitions should be revisited to reflect the fact that current target values for stand structure (e.g. density of large-diameter trees and snags) are lower than the threshold values found for the probability of occurrence (Guénette and Villard 2005) or the probability of nesting (Poulin et al. 2008) of several species of forest songbirds. Finally, unless measures are taken, partial harvesting will continue to degrade habitat for many species requiring closed canopy stands, large live trees or dead wood and its cumulative impacts will become even more significant as we enter second or third partial-cut rotations. Studies such as those summarised here offer options to balance economic values and biodiversity conservation, and they provide excellent opportunities to model trade-offs.

Acknowledgements

I want to thank my graduate students Émilie D'Astous, Jean-Sébastien Guénette, Samuel Haché, Jean-François Poulin and Stéphane Thériault, as well as my research collaborators and partners, without whom none of this research would have been possible. This research was supported by a NSERC Discovery Grant, as well as grants from the New Brunswick Wildlife Trust Fund, the New Brunswick Innovation Foundation, and graduate scholarships from NSERC and J.D. Irving Ltd.

Biography

Marc-André Villard teaches Biology at Université de Moncton. He held a Canada Research Chair in Landscape Conservation from 2001 to 2010. His research focuses on processes underlying species response to landscape change, with a focus on birds in managed forests.

Dr Villard frequently works in partnership with forestry companies to develop field experiments and to recommend practices compatible with biodiversity conservation. He co-chaired the Ecology and Evolution Evaluation Committee at NSERC (Natural Sciences and Engineering Research Council of Canada), he is a Fellow of the American Ornithologists' Union, co-editor of the journal *Avian Conservation and Ecology*, and he co-edited a book entitled *Setting Conservation Targets for Managed Forest Landscapes*.

References

Andrén H (1992) Corvid density and nest predation in relation to landscape fragmentation – a landscape perspective. *Ecology* **73**, 794–804.

Bélisle MM and Desrochers A (2002) Gap-crossing decisions by forest birds: an empirical basis for parameterizing spatially-explicit, individual-based models. *Landscape Ecology* 17, 219–231.

Betts MG and Villard M-A (2009) Landscape thresholds in species occurrence as quantitative targets in forest management: generality in space and time? In *Setting Conservation Targets for Managed Forest Landscapes*. (Eds M-A Villard and BG Jonsson) pp. 185–206. Cambridge University Press, Cambridge, UK.

Betts MG, Forbes GJ and Diamond AW (2007) Thresholds in songbird occurrence in relation to landscape structure. *Conservation Biology* **21**:1046–1058.

Carignan V and Villard M-A (2002) Effects of variations in micro-mammal abundance on artificial nest predation in conifer plantations and adjoining deciduous forests. *Forest Ecology and Management* **15**, 255–265.

Edman M, Eriksson A-M and Villard M-A (2008) Effects of selection cutting on the abundance and fertility of indicator lichens *Lobaria pulmonaria* and *L. quercizans*. *Journal of Applied Ecology* **45**, 26–33.

Erdle T (2008) 'Management alternatives for New Brunswick's public forest'. Internal report, 108 pages.

Gillies CC and St Clair CCS (2008) Riparian corridors enhance movement of a forest specialist bird in fragmented tropical forest. *Proceedings of the National Academy of Sciences* **105**, 19774–19779.

Guénette J-S and Villard M-A (2005) Thresholds in forest bird response to habitat alteration as quantitative targets for conservation. *Conservation Biology* **19**, 1168–1180.

Haché S and Villard M-A (2010) Age-specific response of a migratory bird to an experimental alteration of its habitat. *Journal of Animal Ecology* **79**, 897–905.

Halvorson CH and Engeman RM (1983) Survival analysis for a red squirrel population. *Journal of Mammalogy* **64**, 332–336.

MacLean DA, Hemens B, Higdon J, Etheridge D, Wagner RG, Porter KB, Hagan JM and Scott JM (2002) Using analyses of natural and human-caused forest disturbance on the JD Irving Ltd Black Brook District to inform forest and biodiversity management. In *Advances in Forest Management: From Knowledge to Practice*. (Ed. TS Veeman) pp. 99–104. Sustainable Forest Management Network, Edmonton, Alberta.

NB-DNR (2005) 'Habitat definitions for old-forest vertebrates in New Brunswick'. New Brunswick Department of Natural Resources, Fredericton, New Brunswick.

Pärt T, Arlt D and Villard M-A (2007) Empirical evidence for ecological traps: a two-step model focusing on individual decisions. *Journal of Ornithology* **148**, S327–S332.

Pérot A and Villard M-A (2009) Putting density back into the habitat quality equation: a case study on an open-nesting forest bird. *Conservation Biology* **23**, 1550–1557.

Poulin J-F and Villard M-A (2011) Edge effect and matrix influence on the nest survival of an old forest specialist, the Brown Creeper (*Certhia americana*). *Landscape Ecology* **26**, 911–922.

Poulin J-F, Villard M-A and Haché S (2010) Short-term demographic response of the brown creeper to experimental selection harvesting. *Ecoscience* **17**, 27–40.

Poulin J-F, Villard M-A, Edman M, Goulet P and Eriksson A-M (2008) Thresholds in nesting habitat requirements of an old forest specialist as conservation targets. *Biological Conservation* **141**, 1129–1137.

Pykälä J (2004) Effects of new forestry practices on rare epiphytic macrolichens. *Conservation Biology* **18**, 831–838.

Robertson OJ and Radford JQ (2009) Gap-crossing decisions of forest birds in a fragmented landscape. *Austral Ecology* **34**, 435–446.

Robertson BA and Hutto RL (2007) Is selectively harvested forest an ecological trap for Olive-sided Flycatchers? *Condor* **109**, 109–121.

Schmidt KA, Rush SA and Ostfeld RS (2008) Wood Thrush nest success and post-fledging survival across a temporal pulse of small mammal abundance in an oak forest. *Journal of Animal Ecology* **77**, 830–837.

Vanderwel MC, Malcolm JR and Mills SC (2007) A meta-analysis of bird responses to uniform partial harvesting across North America. *Conservation Biology* **21**, 1230–1240.

11 LAND USE INTENSIFICATION IMPACTS ON BIODIVERSITY IN THE MALLEE/WHEAT LANDSCAPE OF CENTRAL NSW

Saul A. Cunningham, David H. Duncan and Don A. Driscoll

Lesson #1. Field research exposes a moment in time, but landscape transformation has a long-term dynamic.

Lesson #2. Extinction debt will be important in recently developed production landscapes.

Lesson #3. In some landscapes the agricultural matrix plays little role in supporting biodiversity.

Lesson #4. Biodiversity losses can be so extreme that it challenges ideas about what useful conservation goals might be.

Lesson #5. Some elements of biodiversity show surprising signs of resilience to extreme fragmentation.

Lesson #6. In heavily cleared landscapes small remnants have heightened value, as a resource to be protected and as the foundation for improving biodiversity outcomes in the future.

Lesson #7. Managers of both the native and agricultural biota must be responsive to pulses of critical resources.

Lesson #8. Future land use scenarios are constrained by the productive potential of the land.

Introduction

Large-scale cereal cropping is one of the most widespread intensive agricultural practices in the world. Central New South Wales (Australia) is typical of the cereal cropping land-scape in south-eastern Australia, where extensive clearing of the original woodland vege-tation enabled dryland cereal cropping (especially wheat) across most of the plain, with livestock (primarily sheep) usually part of the land use rotation. The level of land clearing in the cereal growing regions of Australia (SE and SW) is such that it is recognised as the

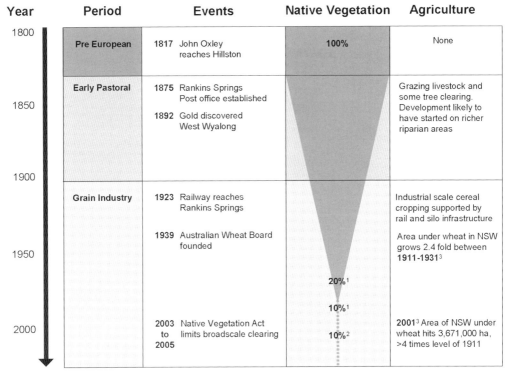

Year	Period	Events	Native Vegetation	Agriculture
1800	Pre European	1817 John Oxley reaches Hillston	100%	None
1850	Early Pastoral	1875 Rankins Springs Post office established

1892 Gold discovered West Wyalong | | Grazing livestock and some tree clearing. Development likely to have started on richer riparian areas |
| 1900 | | | | |
| 1950 | Grain Industry | 1923 Railway reaches Rankins Springs

1939 Australian Wheat Board founded | 20%[1]

10%[1] | Industrial scale cereal cropping supported by rail and silo infrastructure

Area under wheat in NSW grows 2.4 fold between 1911-1931[3] |
| 2000 | | 2003 Native Vegetation Act to   limits broadscale clearing 2005 | 10%[2] | 2001[3] Area of NSW under wheat hits 3,671,000 ha, >4 times level of 1911 |

1 Sivertsen and Metcalfe (1995)
2 Driscoll (2005)
3 Australian Bureau of Statistics (2004)

Figure 11.1: Summary of the history of agricultural development in the focal study landscape (sources: Sivertsen and Metcalfe 1995; Driscoll 2005; Australian Bureau of Statistics 2004). Land-clearing data are only available for a few points of time from the 1970s, but data on the area of wheat in NSW are likely to provide a surrogate (inverse) for the temporal pattern of land clearing, assuming that this landscape approximately reflected the statewide trends.

extreme even in global analyses (Matthews 1983; Hoekstra *et al.* 2005). Since 1997 we have conducted a series of studies exploring the influence of the fragmented habitat structure on metapopulations, species richness of functional groups of fauna, and plant–pollinator interactions in the remnant woodlands of this landscape. These studies have been concentrated in a specifically defined landscape within 50 km radius from 33°54′S, 146°18′E, where the dominant woodland type is mallee vegetation, and our insights have been further enriched by work in similar landscapes in South Australia and Victoria.

The plains in our focal region are mostly 150–200 m ASL, and receive mean annual rainfall of approximately 400 mm, but the pattern is erratic with large inter-annual variation. The landscape has a short (<200 yrs) history of agricultural development (see Figure 11.1). In 1817 the lowland plains were still dominated by mallee woodlands and associated plant communities. In the following century agriculture was introduced, based on grazing livestock and incremental clearing of native vegetation. In the second century (1917–present) the massive growth of cereal cropping transformed the landscape such that there is now less than 10% of native vegetation remaining on the plains (Figure 11.1; Sivertsen and Metcalfe 1995; Driscoll 2005). A burst of land clearing was experienced in the wheat belt as recently as the 1970s, when >50% of the remaining remnant vegetation on the plain was cleared (Sivertsen 1994).

What does intensification mean in the context of this landscape? As we describe below, the landscape we focus on in this chapter is extensively cleared, and those cleared elements are typically intensively managed. Whether this landscape continues to be subject to intensification is uncertain. Given tight controls on land clearing, it is likely that agriculture has already peaked in terms of spatial extent. However, sheep grazing is in long-term decline, and continuous cropping is now possible; thus, even more intensive use of some proportion of the production areas is likely (Duncan *et al.* 2008).

Lessons

1. Field research exposes a moment in time, but landscape transformation has a long-term dynamic

Our field studies describe patterns and processes observed in a series of relatively short-term projects beginning in the late 1990s. Thus, we commenced research once conversion of the mallee landscape to a predominantly cropped landscape was complete, and had been so for a few decades. Therefore, insights into the temporal course of land use intensification require the methods of a historian as much as an ecologist. In the absence of long-term studies we resort to historical records, and methods such as 'space for time' substitution. While our studies were conceived to provide insight into the consequences of fragmentation and other features of landscape modification, the perspective of *land use intensification* comes as the paradigm laid over existing ideas and data. To a greater or lesser degree this will often be the true history of data for any landscape study – as purpose-designed studies of land use intensification are rare.

The conversion of intact mallee landscapes to an extensively cleared and intensively managed landscape occurred swiftly and recently (see Figure 11.1; Duncan and Dorrough 2009). The resultant landscape is strongly polarised into contrasting land uses, with engineered surfaces for cereal production interspersed with native elements that often retain considerable native structural and compositional diversity. These native habitat elements occur most commonly on the plains as narrow linear strips along road and rail lines, and infrequently as small or large reserves. Native habitats also remain on most rocky ridges.

Some of the native remnants are exposed to frequent disturbance such as grazing livestock and earthworks (e.g. roadsides, stock reserves), some from grazing and forestry (e.g. state forests). Even the remnants in nature reserves are chronically exposed to fertiliser run-off and fallout (e.g. Duncan *et al.* 2008).

To understand the likely future trajectory of this landscape one needs to appreciate that many of the landscape changes have happened rapidly relative to the life histories of many of the organisms affected. For example, seed production by the plant species *Eremophila glabra*, *Acacia brachybotrya* and *Dianella revoluta* is lower in linear remnants than in larger reserves (Cunningham 2000a,b; Cunningham and Duncan 2001; Elliot 2010). However, there is no evidence yet that this has affected recruitment. Given that recruitment events are likely to be extremely episodic, driven by temporally pulsed phenomena such as rare wet years and fires (see lesson 7), we do not know when population level effects will be expressed, if at all.

2. Extinction debt will be important in recently developed production landscapes

While many mammals in this landscape have already become regionally extinct (Bilby, Western Quoll, Brush Tail Bettong, Eastern Hare Wallaby, Bridle Nail Tail Wallaby, Gould's Mouse, Koala), the Mallee Fowl – one of the archetypal species of this landscape – are still in the process of contracting in distribution (Benshemesh 2007), so that while they were recently known from some of the small reserves, they are now found only in the largest patches of remnant vegetation. This contraction continues even while the rate of land clearing has stabilised, albeit at a level with only a small percentage (<10%) of remnant cover.

Among the reptiles we studied, two species appeared to be locally extinct, and a further five species had declined or were declining from linear remnants, leaving remaining populations isolated in reserves. A further nine reptile species had patchy distributions through the strips and reserves (Driscoll 2004). These species appeared to be vulnerable to extinction, with the risk that isolated populations may die out over coming decades, with no possibility of re-colonisation (Driscoll 2004). A similar picture emerged for the beetles in these landscapes. Driscoll and Weir (2005) estimated that 21% of the beetle fauna may be at risk of decline over coming decades. Driscoll (2004) recommended that widespread revegetation would be necessary to avoid paying back the extinction debt, representing one possible future scenario for these mallee landscapes. This revegetation would have to occur largely on the privately held farmland that dominates land tenure on the plain.

3. In some landscapes the agricultural matrix plays little role in supporting biodiversity

Our study landscape is quite unlike those known from many other parts of the world where agricultural land management can be seen as a valuable tool in the conservation of certain species or communities (e.g. agroforestry plantations of coffee or cacao, see chapters 14 (Vandermeer and Perfecto) and 2 (Tscharntke *et al.*)). In our landscape the contrast between the dominant agricultural form (a wheat field) and a remnant mallee is stark in both structure and composition. Field sizes are large (e.g. 4–6 km^2) and natural features in fields (e.g. scattered trees) are rare, providing little scope for connectivity through the matrix, except through linear remnants (see lesson 6). Whereas in some fine-grained heterogeneous agricultural landscapes different agricultural land uses can be viewed as offering a dynamic range of ecological opportunities (e.g. mass-flowering supporting native pollinators, Westphal *et al.* 2003), in this landscape the current agricultural land uses have led to large scale homogenisation and simplification.

We know of no native species of conservation concern that rely on the cropping matrix in this landscape. Extensive surveys for reptiles in the interior of cropped or grazed areas found only two individual blind snakes (compared to 1128 individuals and 29 species in remnants, Driscoll 2004, and unpublished data). Seventeen of 116 beetle species were collected more frequently in the fields than remnants (Driscoll and Weir 2005), but there is no evidence that they depend on this habitat for their regional persistence. No endemic flora appears to be dependent on agriculture.

In short, the simplistic view of habitat versus non-habitat is appropriate for most species in the current landscape, to the detriment of biodiversity. This could change if there were a significant move away from cropping and toward extensive perennial grazing as a land use (see Duncan and Dorrough 2009 and Bryan *et al.* 2011 for discussion). Native habitat value of *Atriplex* (saltbush) on farms has been demonstrated (Seddon *et al.* 2009; Collard *et al.* 2011) and Smith (2009) also demonstrated how oil mallee plantations can provide habitat value.

Although there were fewer species, beetles were twice as abundant in fields as in remnants, and this likely reflects substantial biological activity (Driscoll and Weir 2005). These field specialists included scavengers, omnivores and predators, implying that beetles contribute substantially to nutrient and energy flows in fields. One direct biological service that was observed in the field involved the predatory carabid beetle *Calosoma schayeri*. Colloquially known as the stink beetle for its odorous defence mechanism, this beetle achieved enormous densities as it preyed upon lepidopteran larvae that in turn were feeding on a nitrogen-fixing crop (Driscoll, unpublished). We expect that without the service provided by this predator less nitrogen would have been fixed through the rotating-crop approach.

Our research on pollinators in this landscape highlighted native bees that are capable of crossing into wheat fields in search of resources, and pollinating birds that transport pollen many kilometres from patch to patch. However, none of the agricultural crops currently grown in this landscape benefit from insect or bird pollination, and an ecosystem service view will, in this respect, provide no incentive to farmers.

4. Biodiversity losses can be so extreme that it challenges ideas about what useful conservation goals might be

Investment in conservation is generally applied in triage mode, where resources are focused on cases where the urgency and likelihood of ecological recovery are greatest (Hobbs and Kristjanson 2003). With this view, landscapes such as ours might generally be overlooked (Duncan and Dorrough 2009). There are very few remnant patches on the plains, and the linear fragments that criss-cross the landscape are so thin or degraded that they may not be considered by many conservation planners to meet the criteria for conservation investment (but see lesson 6). Many species of conservation concern are already regionally lost and other species of interest are known to occur outside of the study landscape, especially in areas where mallee clearing has been less extensive. So what then become sensible conservation goals for such a transformed landscape?

One approach is to prioritise populations rather than species. Driscoll and Hardy (2005) found that populations of the dragon *Amphibolurus nobbi* in the farming landscape were genetically distinct from those in uncleared mallee to the north of our study region, and satisfy the criteria for an evolutionarily significant unit (Moritz 1994). This species has declined in the farmed landscape, and was among species that Driscoll (2004) regarded as at most risk of extinction. So loss of species from our study region may not represent a global extinction, but may represent an irreversible loss of evolutionary potential.

A more conservative goal would be to secure a landscape that will support at least those species that are resilient to the extreme land clearing. This would require a better understanding of the processes that support persistence in such an apparently fragmented state.

5. Some elements of biodiversity show surprising signs of resilience to extreme fragmentation

While many species are in decline, many other species are predicted to survive in this landscape. Fifteen per cent of beetle species were more abundant in fields than strips or reserves. While some of these species may depend on remnant vegetation to complete part of their life cycle, they are surviving well in the current mix of remnants and extensive fields (Driscoll and Weir 2005).

One skink species, *Lerista punctatovittata*, was three times more abundant in roadside remnants than in larger remnant patches, possibly linked to increased nutrient run-on beside dusty roads (Driscoll 2004).

Among the plants we have studied, three species showed significantly lower fruit/flower ratios when in linear fragments, but this phenomenon was sometimes counterbalanced by higher flowering intensity or lower fruit predation (Cunningham 2000b). Although *Eremophila glabra* has lower pollen receipt and lower seed set in linear strips (Cunningham 2000a; Cunningham and Duncan 2001; Elliott 2010) gene flow through pollination was remarkably resistant to fragmentation, with 70–80% of progeny being sired by pollen from outside the local patch, regardless of whether the patch was in a large remnant or narrow linear strip. This extraordinary pattern must be due to frequent long distance pollen dispersal by the pollinating birds (honeyeaters) combined with strong genetic selection mechanisms in the plant.

Visitation by native bees also seemed robust to fragmentation. For *Dianella revoluta* we found that isolated flowers in linear strips still received pollinating visits. Where there was a shortage of pollen, it was due not to a lack of visits but rather a reduced pollen supply due to reduced mate density (Duncan *et al.* 2004a,b). Most of the native bees are ground nesting and probably have ample breeding opportunities even outside large remnant patches, in road verges or field margins.

6. In heavily cleared landscapes small remnants have heightened value, as a resource to be protected and as the foundation for improving biodiversity outcomes in the future

We found that the fine cobweb of linear connections on the plains landscape was used by a wide range of native fauna. Further, data on gene flow through long distance pollen transport, combined with observations of birds, suggest that the linear remnants can provide important connectivity for evolution and persistence of some plant species (Elliot 2010).

Theoretical ecology continues to emphasise the potential for dispersal to influence local community composition. High population growth rates coupled with dispersal have the potential to maintain populations across the intensively farmed landscapes through source-sink dynamics (Pulliam 1988). The hypothesis that linear remnants are particularly important in providing resilience to the effects of fragmentation needs to be tested. In the meantime, we see degradation of remaining remnants by livestock grazing, and piecemeal damage to roadside remnants through roadworks and fence works. Reducing this loss is of paramount importance. Unfortunately the nature reserve selection paradigm usually places a low priority on landscape elements like these (Margules and Pressey 2000). Degradation of these remnants has led to substantial habitat loss and increased fragmentation, despite the appearance of a relatively well-connected landscape (Driscoll 2004; Driscoll and Weir 2005; Driscoll and Hardy 2005).

In contrast, much larger areas of remnant vegetation in this landscape occur on or around the rocky ridges. While the plant communities of the ridges are distinct from those on the plain, some organisms could use the ridge top environment in a way that helps their persistence, either by using the habitat as a route for dispersal, or as a secondary habitat for occasional breeding or resource acquisition. Our research strategies to date have not allowed us to explore this possibility. Future research and conservation planning should pay greater attention to the contribution the ridge remnants might make to sustaining plains-based biodiversity.

7. Managers of both the native and agricultural biota must be responsive to pulses of critical resources

The Australian climate is characterised by extreme inter-annual variation in rainfall, driven in part by the El Niño/Southern Oscillation phenomenon. This is known to have great impacts on natural phenomena, such as flowering intensity, vegetation structure, population dynamics of short-lived animals and movement dynamics of far-ranging animals. Equally, in wet years farmers will crop more extensively and in dry years crop less, and livestock are moved around the landscape in search of grazing, including public land. These phenomena can be difficult to study because the events are unpredictable and cannot be planned for by researchers. However, the reality may be that long-term population dynamics are being driven by occasional critical events.

As well as understanding the intermittent nature of recruitment and dispersal events, managing threats requires attention to the dry years. Grazing of public land during drought is a feature of much of the Australian landscape, known to pose threats to biodiversity. Grazing had a big impact on three reptile species, including complete elimination of one species in grazed sites (Driscoll 2004). When remnant vegetation is grazed during drought, the impacts on native species could be substantially magnified as the few resources available during drought are further reduced (Duncan *et al.* 2008; Retzer *et al.* 2006).

8. Future land use scenarios are constrained by the productive potential of the land

Cropping in our study landscape is lower yielding than further east where average rainfall is greater. Drought has been particularly frequent in recent decades and many models predict increasingly intense droughts associated with climate change (http://www.csiro.au/ozclim/home.do). Sheep grazing has been in long-term decline with poor wool prices. As is the case in much of rural Australia, the average age of the farming community is increasing, and young people are often choosing to live elsewhere. Given these trends it is difficult to anticipate what the future land use trends will be in this landscape, or where the investment might come from to support better biodiversity management.

As Australia develops a market for carbon sequestration, this might open the door to some land use change. Bryan *et al.* (2011) envisage mallee landscapes extensively replanted under carbon market conditions, as well as biodiversity and 'conservation farming' scenarios. However, the potential to fix carbon is limited by rainfall, just as it limits yield of the annual crops: whether or not carbon sequestration becomes a profitable land use in this landscape remains to be seen. The spread of more conservation oriented farming strategies requires either that they are proven to be profitable in their own right, or that economic incentives are provided to foster adoption.

Conclusions

We have focused on a landscape where the majority of the native vegetation has been cleared and replaced with large-scale cereal cropping, with substantial negative impacts on biodiversity. The future for biodiversity in this landscape continues to be threatened by lag effects (like extinction debt), especially given the relatively recent history of development. While we can conceive of land use changes that would benefit biodiversity, optimism for the future is constrained by the fact that the dominant land use (cereal cropping) seems to offer little potential for conservation of the native biodiversity, and the economic prospects for alternative land uses remain uncertain.

While these insights have their origin in a very local perspective, there are other landscapes around the world where many of the circumstances are similar. Cereal cropping accounts for approximately 15% of all agricultural land (FAO 2011) and this land use often occurs in the most heavily cleared landscapes (Matthews 1983). Because cereals are the mainstay of the human diet, current projections expect increasing growth to drive ongoing intensification of this production system (Cassman 1999).

Although biodiversity conservation in these landscapes is difficult, there are nevertheless elements of biodiversity that persist in the scattered remnant vegetation. Theory suggests that connectivity among sub-populations is critical to maintaining these species in the landscape, and in our focal landscape that connectivity might be provided by the fine network of linear remnants. More work is needed to establish the role of remnant condition and configuration for species in this landscape, but in the meantime these remnants are exposed to ongoing threats. We suggest that in this landscape, and others like it where clearing of native vegetation has been so extensive, the scattered small remnants deserve to be protected because of their potential to support significant biodiversity, and because they are vulnerable to piecemeal damage.

Acknowledgements

We acknowledge the contributions of the many other researchers who have contributed to our understanding of the ecology of the mallee landscape, including Carole Elliott, Linda Broadhurst, Andrew Young, Tom Weir, Josh Dorrough, Chris Hardy and Adrienne Nicotra.

Biographies

Saul Cunningham studied habitat fragmentation impacts on plants and pollination in this landscape, commencing in 1997. He has maintained a connection with the area through supervision of students who have extended the project in the years since. Since 1999 he has been a research scientist with CSIRO. The major themes in his work are understanding pollination processes in landscapes with a recent focus on crop pollination and its significance to agriculture, and understanding how biodiversity responds to different land management strategies.

David Duncan conducted his PhD research from 2000–2003 on the effects of habitat fragmentation on the pollination success of plants in NSW mallee woodlands. In 2004 David joined the Victorian Department of Sustainability and Environment where he works as a vegetation scientist. He continued to work in mallee systems in Victoria from 2004–2007, conducting broad-scale surveys of ecological condition of remnant woodlands and an investigation into the effect of removing stock grazing pressure from degraded sites. David's work continues to focus on modelling and measuring how vegetation structure and composition of sites and landscapes respond to current and historical land management practices.

Don Driscoll has studied the consequences of habitat loss and fragmentation in most Australian states over the past 20 years, with an emphasis on frogs, beetles and reptiles. Some ambitious recent projects include attempting to integrate rational decision-making approaches with fire management, and developing methods for rapid measurement of dispersal. Don is currently a research fellow in the Fenner School of Environment and Society at the ANU. He supervises PhD scholars as they discover how to maintain biodiversity in fragmented landscapes, how to conserve threatened frogs and how population and demographic traits of plants, birds and reptiles mediate fire effects.

References

ABS (2004) *New South Wales Yearbook*. Australian Bureau of Statistics, Canberra.

Benshemesh J (2007) 'National Recovery Plan for Malleefowl *Leipoa ocellata*'. Department of Environment and Heritage, South Australia.

Bryan BA, Crossman ND, King D and Meyer WS (2011) Landscape futures analysis: assessing the impacts of environmental targets under alternative spatial policy options and future scenarios. *Environmental Modelling & Software* **26**, 83–91.

Cassman KG (1999) Ecological intensification of cereal production systems: yield potential, soil quality, and precision agriculture. *Proceedings of the National Academy of Sciences* **86**, 5952–5959.

Collard SJ, Fisher AM and McKenna DJ (2011) Planted saltbush (*Atriplex nummularia*) and its value for birds in farming landscapes of the South Australian Murray Mallee. *Ecological Management & Restoration* **12**, 37–45.

Cunningham SA (2000a) Depressed pollination in habitat fragments causes low fruit set. *Proceedings of the Royal Society B* **267**, 1149–1152.

Cunningham SA (2000b) Effect of habitat fragmentation on the reproductive ecology of four mallee woodland species. *Conservation Biology* **14**, 758–768.

Cunningham SA and Duncan DH (2001) Plant reproduction in a fragmented landscape. In *Tropical Ecosystems: Structure, Diversity and Human Welfare. Proceedings of the International Conference on Tropical Ecosystems.* (Eds KN Ganeshaiah, R Uma Shaanker and KS Bawa) pp. 93–96. Oxford-IBH, New Delhi.

Driscoll DA (2004) Extinction and outbreaks accompany fragmentation of a reptile community. *Ecological Applications* **14**, 220–240.

Driscoll DA (2005) Is the matrix a sea? Habitat specificity in a naturally fragmented landscape. *Ecological Entomology* **30**, 8–16.

Driscoll DA and Hardy CM (2005) Dispersal and phylogeography of the agamid lizard *Amphibolurus nobbi* in fragmented and continuous habitat. *Molecular Ecology* **14**, 1613–1629.

Driscoll DA and Weir T (2005) Beetle responses to habitat fragmentation depend on ecological traits, habitat condition, and remnant size. *Conservation Biology* **19**, 182–194.

Duncan DH and Dorrough JW (2009) Historical and current land use shape landscape restoration options in the Australian wheat and sheep farming zone. *Landscape & Urban Planning* **9**, 124–132.

Duncan DH, Dorrough J, White M and Moxham C (2008) Blowing in the wind? Nutrient enrichment of remnant woodlands in an agricultural landscape. *Landscape Ecology* **23**, 107–119.

Duncan DH, Nicotra AB and Cunningham SA (2004a) High self-pollen transfer and low fruit set in buzz-pollinated *Dianella revoluta* (Phormiaceae). *Australian Journal of Botany* **52**, 185–193.

Duncan DH, Nicotra AB, Wood JT and Cunningham SA (2004b) Plant isolation reduces outcross pollen receipt in a partially self-compatible herb. *Journal of Ecology* **92**, 977–985.

Driscoll DA and Weir T (2005) Beetle responses to habitat fragmentation depend on ecological traits, habitat condition, and remnant size. *Conservation Biology* **19**, 182–194.

Elliott C (2010) Patterns and processes: ecological and genetic function of fragmented Emu bush (*Eremophila glabra* ssp. *glabra*) populations. Ph.D. Thesis, The Australian National University, Canberra.

FAO (2011) FAOSTAT website, accessed June 2011 (http://faostat.fao.org/site/339/default. aspx)

Hobbs R and Kristjanson LJ (2003) Triage: how do we prioritize health care for landscapes? *Ecological Management & Restoration* **4**, S39–S45.

Hoekstra JM, Boucher TM, Ricketts TH and Roberts C (2005) Confronting a biome crisis: global disparities of habitat loss and protection. *Ecology Letters* **8**, 23–29.

Margules CR and Pressey RL (2000) Systematic conservation planning. *Nature* **405**, 243–253.

Matthews E (1983) Global vegetation and land use: new high-resolution data bases for climate studies. *Journal of Climate and Applied Meteorology* **22**, 474–487.

Moritz C (1994) Defining 'Evolutionarily Significant Units' for conservation. *Trends in Ecology and Evolution* **9**, 373–375.

Pulliam HR (1988) Sources, sinks and population regulation. *American Naturalist* **132**, 652–661.

Retzer V, Nadrowski K and Miehe G (2006) Variation of precipitation and its effect on phytomass production and consumption by livestock and large wild herbivores along an altitudinal gradient during a drought, South Gobi. Mongolia. *Journal of Arid Environments* **66**, 135–150.

Seddon J, Doyle S, Bourne M, MacCallum R and Briggs S (2009) Biodiversity benefits of alley farming with old man saltbush in central western New South Wales. *Animal Production Science* **49**, 860–868.

Sivertsen D (1994) The native vegetation crisis in the wheat belt of NSW. *Search* **24**, 5–8.

Sivertsen D and Metcalfe L (1995) Natural vegetation of the southern wheat belt. *Cunninghamia* **4**, 103–128.

Smith FP (2009) Assessing the habitat quality of oil mallees and other planted farmland vegetation with reference to natural woodland. *Ecological Management & Restoration* **10**, 217–227.

Westphal C, Steffan-Dewenter I and Tscharntke T (2003) Mass flowering crops enhance pollinator densities at landscape scale. *Ecology Letters* **6**, 961–965.

12 ECOLOGICAL RESPONSES OF AUSTRALIAN GRASSY WOODLAND AND SHRUBLAND ECOSYSTEMS TO AGRICULTURAL INTENSIFICATION: LESSONS FROM LONG-TERM, MULTI-SPECIES, MULTI-BIOME STUDIES

Andrew Young and Linda Broadhurst

Lesson #1. Multi-disciplinary studies of species and ecosystem responses to agricultural intensification using a range of complementary approaches that fill in the whole ecological story are by far the most efficient and informative.

Lesson #2. Pattern analysis is informative, but accurate quantitative prediction about biodiversity trajectories that can be used to develop and evaluate alternative management options generally depends on understanding process, and this requires detailed longitudinal studies.

Lesson #3. High levels of biodiversity and ecological function can persist in intensively managed landscapes.

Lesson #4. Generalities of ecological response to land use intensification exist across species and even biomes and these can form key management guidelines for achieving conservation.

Lesson #5. Interactions between related ecological processes often result in significant non-linear dynamics that drive systems and, importantly, create thresholds of stability, viability and persistence.

Lesson #6. Connectivity is real for plants and as a result there is often a disconnect between the biologically appropriate scale for management and the enterprise scales of land ownership which is where land management decisions are made and implemented.

Lesson #7. Having 'flagship' species or ecosystems that land holders can recognise and identify with can be a powerful tool for motivating conservation focused land use activities, especially if it is possible to identify multiple values to native species beyond direct biodiversity conservation.

Introduction

Over the past 15 years we have worked in grassland, grassy woodland and mallee woodland communities in south-eastern Australia trying to understand how native plants have responded to historical land use changes, how they are coping with current patterns of land use and their future capacity to persist and evolve. Over the last 100 years this region has undergone extensive conversion of native plant communities to pastoral sheep farming enterprises or cropping, primarily with wheat and more recently canola, in combination with legume rotation crops. The last 5–10 years have seen a second significant shift in land use, especially in the High Rainfall Zone (>roughly greater than 800 mm), from sheep grazing to mixed cropping often using grazing wheats and canolas. In terms of conserving Australia's unique biodiversity, effective management of these fragmented native ecosystems in a production context is crucial as they encompass a large number of endemic plant and animal species and communities, several hundred of which are protected under the Commonwealth Government's *Environment Protection and Biodiversity Conservation Act* 1999 (EPBC Act 1999). Often these species and ecosystems are not adequately represented in the National Parks and Reserves Network or in state protected areas, so management for long-term conservation must occur within the context of private ownership and ongoing agricultural productivity.

The broad aim of our research in these now thoroughly intermixed native and agricultural landscapes has been to provide the biological knowledge required to underpin the development of landscape-scale management concepts and, importantly, enterprise-level land management practices, that maximise the likelihood of native species persistence. Throughout, our work has had a strong focus on plant ecological and genetic responses to habitat loss and fragmentation as well as a bias towards population biology and evolutionary ecology. The research approach has included field-based ecological monitoring, molecular ecological analysis and glasshouse and field-based experiments, as well as a significant component of demographic and genetic modelling. Several long-term studies of single species (e.g. the grassland herb *Rutidosis leptorrhynchoides* and the mallee shrub *Eremophila glabra*) have been used to derive deep quantitative understanding of biological process. We have attempted to complement this with a larger number of shorter-term studies of a range of associated species with similar and contrasting life-history traits. Data from these have been used to try and understand generalities of effect.

In this chapter we describe three broad kinds of lessons that we believe our studies suggest to be true regarding the science-based management of biodiversity in agricultural landscapes, though these are not entirely discrete: (1) lessons about how to study or know things; (2) generalities about observed biological effects of agricultural intensification; and (3) knowledge about the challenges of translating biological understanding into effective conservation outcomes and the opportunities presented by emerging agricultural practices and technologies.

Lessons

1. Multi-disciplinary studies of species and system responses to agricultural intensification that use a range of complementary approaches to fill in the whole ecological story are the most efficient and informative

When investigating ecological responses to habitat change, our experience consistently points to the value of multi-disciplinary approaches which we believe provide greater explanatory depth of understanding about how biological processes are changing at population, species, landscape and ecosystem levels. By integrating complementary approaches (e.g. field monitoring, manipulative experiments, molecular ecological analyses and simulation modelling), that can isolate the dynamics of particular ecological processes (e.g. mating, dispersal, recruitment and reproduction), it is possible to identify factors that limit demography and population viability. Shorter-term studies of these processes can then be undertaken on a broader range of species to assess the generality of observed effects and so extrapolate management lessons to other species with similar life history traits. This is a powerful two-tiered model for rapidly building a reasonably detailed whole-of-ecosystem understanding without engaging in oversimplification that can rob studies of genuine predictive power.

2. Pattern analysis can be informative but accurate quantitative prediction that can be used to develop and evaluate alternative management options generally depends on understanding process and this requires longitudinal studies

Understanding whole of life-cycle responses to habitat change can be critical in predicting medium-term demographic trajectories and appropriate restoration actions. For example, recruitment failure is a common constraint to long-term persistence of small populations of perennial herbs in the temperate native grasslands of south-eastern Australia. However, this can result from different causes with different solutions. In the self-compatible herb *Swainsona recta* (Fabaceae) it is due to high levels of inbreeding reducing seed germination and seedling establishment (Buza *et al.* 2000). In contrast, for the self-incompatible plant *Rutidosis leptorrhynchoides* (Asteraceae), poor recruitment results from low seed production because of limited mate availability due to low diversity of incompatibility genotypes (Young *et al.* 2000). This differential response mediated by breeding system means that recovery strategies to promote population viability are different. For *S. recta*, simply increasing local populations' sizes by sourcing more individuals locally will alleviate the inbreeding and restore fitness. For *R. leptorrhynchoides*, however, increasing local population size is of no help. New genetic material from other populations with different S alleles must be imported to restore mate availability (Pickup and Young 2007).

3. High levels of biodiversity and ecological function can persist in intensively managed landscapes

Emu bush, *Eremophila glabra* (Myoporaceae), is a bird-pollinated species from the intensively cropped mallee landscapes of western NSW that continues to persist despite being confined to linear habitat strips along crop field boundaries. Two life history traits, both of which promote reproductive assurance, are responsible for this ongoing persistence. The first is the highly vagile nature of the local honeyeater species which are responsible for pollination. Bird movement data and genetic analysis of seed paternity show that honeyeaters will travel large distances (~15 km) if sufficient reward is available (Elliott 2010). The second factor is the plant's self-incompatible breeding system that favours immigrant pollen grains in fertilisation events as these are more likely than local pollen sources to have different mating genotypes. When combined, these factors promote inter-population

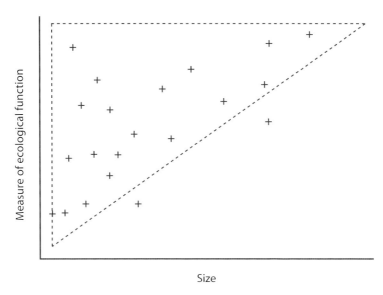

Figure 12.1: Large populations or sites consistently exhibit high ecological functionality, smaller populations have lower and more variable function

connectivity which means that *E. glabra* populations function ecologically at a much greater scale than the individual patch populations they appear to be. This also demonstrates the dependence of species, especially plants, on ecological interactions with other species – in this instance pollinators. Such results highlight how second order effects of agricultural intensity on interacting partner organisms can either positively or negatively influence ecological outcomes.

4. Generalities of ecological response to land use intensification exist across species and even biomes and these can form key management guidelines

Population/habitat patch size is still perhaps the most useful predictor of ecological function for plants. Larger populations consistently harbour high genetic diversity, have low inbreeding, good pollinator service, high fecundity and substantial recruitment. Small populations are on average worse with regard to these kinds of ecological and genetic parameter values and measures of demographic viability – very importantly they are also much more variable. This variation is because initial losses of individuals, genetic variation, pollinators etc. due to sampling effects at the time of habitat fragmentation sets up what can be very different starting conditions for small populations. In contrast, populations that remain large probably all start off their post-fragmentation ecological trajectories in much the same condition. This higher variation is also likely to be at least partly driven by the fact that small populations are more influenced by their spatial and ecological context within the landscape than large ones. Overall, the result is a commonly observed triangle-shaped relationship for a range of ecological response variables that are associated with ecological function, population viability and persistence (as shown in Figure 12.1). This relationship provides practical guidance with respect to prioritising sites for conservation as big ones are best, but carefully chosen small ones with high ecological

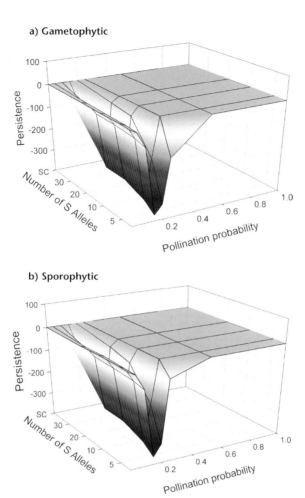

Figure 12.2: Non-additive effects of pollination probability and S allele number on population persistence for: a) gametophytic and b) sporophytic self-incompatibility systems. Deviations from zero represent effects that depart from a simple additive model of joint action. (From Young *et al.* 2012).

function (likely to depend on landscape context with regard to other populations – connectivity) can make significant contributions.

5. Interactions between processes often result in important non-linear dynamics that drive systems and, importantly, create thresholds of stability, viability and persistence

Our detailed studies of interacting population processes in 'generalised model' plant species using individual-based simulations strongly suggest that population viability is often determined by dynamic, non-linear interactions between processes whose independent effects appear not to be particularly important in determining demographic outcomes. These interactions can occur over a relatively small parameter space (at least for one of the variables involved) that may not generally be encountered in more ecologically intact systems. Feedbacks between processes can generate very strong thresholds of ecological function for these parameters that would not be predicted assuming additive effects of either process in isolation. A good example of this phenomenon is the interacting effects of pollinator failure and genetic mate limitation in grassland herbs reported by Young *et al.* (2012) which demonstrates that while pollen or S allele-based mate limitation alone are unlikely to limit population viability except when extreme values are reached, their interactions rapidly reduce fecundity and population persistence across a range of values of both parameters that are commonly encountered in populations found in fragmented landscapes (see Figure 12.2a and b). In such cases, simulation modelling can be very helpful in delineating stable states and likely thresholds for use as a basis for conservation planning. They can also identify possible indicator variables that can be used to monitor when populations are reaching these thresholds.

6. Connectivity is real for plants and as a result there is often a disconnect between the biologically appropriate scale for management and the enterprise scales of land ownership which is where land management decisions are really made and implemented

Spatial analysis of mating dynamics that we have undertaken using genetic-based paternity assignment of seed and behavioural observations of pollinators of several bird-pollinated species (e.g. *Grevillea iaspicula* and *Eremophila glabra*) clearly demonstrate that the primary scale of pollination events is generally greater than a single population or even farming enterprise. Indeed it is common for greater than half the observed mating events to result from inter-population pollination events. Recent meta-analysis of patterns of seed paternity in plants across a range of species with different life histories reported by Ashley (2010) show that this is a common phenomenon – that is, inter-population connectivity is very real for plants. This reproductive interdependence of populations on other populations occurring across several farms makes them highly vulnerable to management decisions made in isolation at the enterprise level. This is the most common scale for land management decisions to be implemented. Mechanisms do not generally exist to inform land managers about how their actions will impact on biodiversity beyond their farm. This is a key barrier to the adoption of integrated conservation-focused land use planning in agricultural landscapes that the ecological data tell us is required.

7. Having 'flagship' species or ecosystems that land holders can recognise and identify with can be a powerful tool for motivating conservation focused land use activities, especially if it is possible to identify multiple values to native species beyond direct biodiversity conservation

Biodiversity outcomes are multiplied when functional benefits beyond simplistic species or ecosystem conservation outcomes within an agricultural production system are identified and can help motivate farmers to undertake conservation activities. Some 10 million hectares of box woodlands, including Grassy White Box Woodlands, once existed across south-eastern Australia but these have been substantially reduced since European settlement. Yellow Box (*Eucalyptus melliodora*) forms part of a community with White Box (*Eucalyptus albens*), Blakely's Red Gum (*Eucalyptus blakelyi*) and an associated understorey of native grasses including Kangaroo and Tussock grasses that is now protected by territorial (ACT), state (NSW) and Commonwealth legislation. The community is home to a range of birds and animals including several that are also legislatively protected. Beyond its value as a conservation species, Yellow Box in particular is highly valued for its durable timber qualities, moderate tolerance for saline areas, excellent firewood, shade and wind-breaks for stock, and honey production and is a key species for broad-scale revegetation across Victoria, NSW and south-central Queensland (Marcar and Crawford 2004; Stelling 1998; Windsor 1998). Large trees such as Yellow Box also play an important role in main-taining hydrological balances in agricultural landscapes where the removal of deep-rooted perennials can lead to dryland salinity. Understanding and articulating these additional woodland values to farmers have played an important part in promoting conservation actions for this vegetation community on private land.

Conclusions

Conservation of native plant biodiversity within intensively farmed agricultural land-scapes is both necessary and possible for a range of Australian species and ecosystems. However, successful conservation outcomes can only come from the recognition of a few key principles. First, management must be based on a quantitative understanding of the spatial and temporal dynamics of key ecological processes and this knowledge is best gained from longitudinal, multi-disciplinary, multi-species studies. High levels of plant biodiversity persist and can be conserved in mixed native–agricultural landscapes but, as with animals, maintenance commonly requires whole landscape planning beyond the scale of the individual farming enterprise. Thresholds (size, connectivity) of ecological function are real and they are often generated by interactions between ecological pro-cesses that alone do not appear to be limiting. Combined, this knowledge can be used to develop landscape design and management principles that can substantially improve ecological function and the likelihood of species persistence.

Finally, and very importantly, there is a receptive audience in farmers for an integrated conservation–production management message. Our general experience with both croppers and graziers has been that they are willing to consider and, where possible imple-ment, a range of conservation-focused activities to manage biodiversity on their farms, if they are provided with the necessary information. They are often also prepared to provide resources towards this end. Indeed the adoption of new farming practices in several key regions such as the High Rainfall Zone of the south-east is likely to present a range of new opportunities in this regard. This is because the potential for higher yields and returns from mixed farming systems that utilise the latest grazing wheats and canolas are giving farmers in this region the financial security to invest in undertaking biodiversity-focused land management activities.

Biographies

Andrew Young is Director of the Centre for Australia National Biodiversity Research – a joint biodiversity research initiative between Parks Australia and the CSIRO. He is a plant ecological geneticist whose research focuses on understanding limits to plant population viability and persistence in fragmented agricultural landscapes. He is an Adjunct Professor at The Australian National University.

Linda Broadhurst is a Senior Research Scientist at the CSIRO Division of Plant Industry. She is a plant molecular ecologist with 15 years' experience in using molecular genetic tools and ecological analysis to examine patterns of plant response to habitat loss and disturbance, and how this information can be used to restore Australian native plant systems.

References

Ashley MV (2010) Plant parentage, pollination, and dispersal: How DNA microsatellites have altered the landscape. *Critical Reviews in Plant Science* **29**, 148–161.

Buza L, Young A and Thrall P (2000) Genetic erosion, inbreeding and reduced fitness in fragmented populations of the endangered tetraploid pea *Swainsona recta*. *Biological Conservation* **93**, 177–186.

Elliot C (2010) Patterns and processes: ecological and genetic function of fragmented Emu bush (*Eremophila glabra* ssp. *glabra*) populations. Unpublished PhD Thesis. The Australian National University, Canberra.

Marcar NE and Crawford DF (2004) 'Trees for Saline Landscapes'. RIRDC, Canberra.

Pickup M and Young AG (2008) Population size, self-incompatibility and genetic rescue in diploid and tetraploid races of *Rutidosis leptorrhynchoides* (Asteraceae). *Heredity* **100**, 268–274.

Stelling F (1998) 'South West Slopes Revegetation Guide'. Murray Catchment Management Committee & Department of Land and Water Conservation, Albury.

Windsor D (1998) 'A landscape approach to optimise recruitment of woodland species in an intensive agricultural environment in the Central Tablelands of NSW'. Charles Sturt University, Mitchell, Bathurst.

Young AG, Brown AHD, Murray BG, Thrall PH and Miller C (2000) Genetic erosion, restricted mating and reduced viability in fragmented populations of the endangered grassland herb Rutidosis leptorrhynchoides. In *Genetics, Demography and Viability of Fragmented Populations*. (Eds AG Young and GM Clarke) pp. 335–359. Cambridge University Press, Cambridge.

Young AG, Broadhurst LM and Thrall PH (2012) Non-additive effects of pollen limitation and self-incompatibility reduce plant reproductive success and population viability. *Annals of Botany* **109**, 643–653.

13 LAND USE INTENSIFICATION IN NATURAL FOREST SETTINGS

David Lindenmayer

Lesson #1. Land use intensification has a spatial dimension and a temporal dimension.

Lesson #2. There can, and often will be negative feedbacks between the temporal and spatial dimensions of land use intensification

Lesson #3. There are general principles for forest biodiversity conservation that are a valuable antidote to the problems associated with land use intensification.

Lesson #4. Land sparing in forests needs to be examined in terms of the quality of the land being spared, the kinds of species targeted for conservation and the conceptual landscape model that underpins decision making.

Lesson #5. Many of the problems associated with land use intensification are underpinned by deeper social and economic problems driven by the overcommitment of natural resources.

Lesson #6. There are always ecological surprises – the key is to learn how to better anticipate and detect them so they are not surprises.

Introduction

I have studied, either directly or indirectly, various aspects of the relationships between biodiversity conservation and land use intensification for most of my (25+ year) research career. The lessons I have learned are derived primarily from my long-term empirical studies in native forests and plantations of trees in south-eastern Australia. However, these lessons are also informed by my work in forests in north-western North America, South America and Europe as well as ongoing large-scale studies in the temperate woodlands within the agricultural and cropping-dominated landscapes of south-eastern Australia.

Lessons

1. Land use intensification has a spatial dimension and a temporal dimension

For the purposes of this chapter, land use intensification can be broadly defined as:

> The increased intensity of human use of the land in a given area or the increased number of areas dedicated to a given form of (productive) human land use across a landscape.

Implicit in this definition are both the temporal and spatial dimensions to land use intensification. Natural forests used for wood and paper pulp production provide a good system in which to examine the temporal and spatial dimensions of land use intensification.

First, logging operations can lead to a significant reduction in the stand structural complexity of natural forests and such effects can persist for decades and sometimes even centuries (reviewed by Lindenmayer and Franklin 2002). Repeated logging can result in vegetation structure being cumulatively simplified over time within a given stand (i.e. after recurrent cutting events: Linder and Östlund 1998). These changes can, in turn, have negative effects on many elements of the biota through well-known relationships like niche richness, vertical heterogeneity and species richness (e.g. MacArthur and MacArthur 1961).

Second, logging of many different harvest units dispersed across a wood production landscape can lead to a spatially homogenous forest landscape dominated by young, relatively recently cut stands (Franklin and Forman 1987). This means that a nominal rotation time of, for example, 80 years will result in a highly 'regulated forest' with all stands aged between 1–80 years and little or no old forest exceeding 81 years old (Lindenmayer and Franklin 2002).

Intensification of forestry operations can lead to many structurally simplified stands dominating large areas of homogenous forest. Therefore, problems of the loss of structural complexity within stands can accumulate over large areas of forest. This can lead to homogenous forest landscapes, thereby creating problems for biodiversity at multiple spatial scales (Franklin and Forman 1987). Tackling and/or reversing problems associated with the intensification of forestry operations will require a comprehensive multi-scaled approach underpinned by strategies at the tree, stand, patch and landscape levels (Lindenmayer 2009) – as per lesson 3 below.

Problems with the spatial and temporal effects of land use intensification are now well recognised in places such as Scandinavia (Gustafsson et al. 2010). There, many decades (and even centuries) of forest clearing and forest harvesting have led to the widespread simplification of stand structure (e.g. Linder and Östlund 1998) and corresponding losses in biodiversity. Swedish forest ecologists now take Swedish forest managers to Russia (where the history of forest harvesting is generally shorter and has been less intensive) to show them what Swedish forests used to look like and to highlight the types of structural complexity that need to be restored to conserve a raft of endangered species (Angelstam et al. 2004) (see Figure 13.1). Unfortunately, the problems associated with the impacts of land use intensification in the northern hemisphere forests are now being repeated in many different parts of the world, including Australia. This makes it clear that we must learn lessons

Figure 13.1: Image series associated with stand simplification in logged forests in Northern Europe. A & B. Old-growth stands near Moscow that support large numbers of species categorised as red-listed taxa in Scandinavia. C. Head of forest management for the area depicted in A & B. D. Field workshop in southern Sweden focusing on the importance of increasing stand structural complexity for enhanced forest biodiversity conservation (see text). (Photos by David Lindenmayer)

from other jurisdictions and ensure that mistakes made elsewhere are not repeated in new places.

2. There can, and often will be negative feedbacks between the temporal and spatial dimensions of land use intensification

Lesson 1 above highlighted how problems with land use intensification at relatively small spatial scales can accumulate in space and in time to become problems at larger spatial scales. There also can be important negative feedback arising from land use intensification at different spatial scales that can radically alter landscapes and significantly impair key ecological processes, sometimes extremely rapidly. Major changes in vegetation cover driven by cutting many different stands of forest can significantly alter the frequency, severity and spatial contagion of natural disturbances like bushfires and windstorms (Franklin and Forman 1987; Lindenmayer *et al.* 2009). Moist temperate and tropical forest landscapes subject to widespread logging operations are significantly more likely to burn, and burn at a higher severity, than equivalent unlogged landscapes (Lindenmayer *et al.* 2009; Malhi *et al.* 2009). Such negative feedback has led to the development of a new kind of ecological trap, termed a landscape trap, in the Mountain Ash (*Eucalyptus regnans*) forests of the Central Highlands of Victoria (south-eastern Australia) in which entire

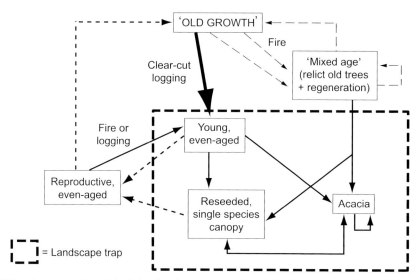

Figure 13.2: Conceptual model of the development of a landscape trap in the Mountain Ash forests of the Central Highlands of Victoria (south-eastern Australia). Redrawn from Lindenmayer *et al.* 2011.

landscapes are trapped in an altered state and are highly unlikely to return to their former functional state (see Figure 13.2). This new type of trap is arising through a combination of logging and fire that favours a regime change (*sensu* Connell and Sousa 1983) from old-growth forest to young fire-prone forests that don't survive to become old growth (Lindenmayer *et al.* 2011).

The potential for landscape traps to develop is likely to be widespread in moist forest ecosystems around the world that are subject to intensive resource management and where bushfire is a major kind of natural disturbance, logging is a widespread form of human disturbance, and fire and logging interact in multiple temporal and spatial feedbacks. In summary, negative feedbacks arising from land use intensification at different spatial scales may result in severely impaired ecological functions of a landscape in an altered state such as reduced suitability for biodiversity, reduced carbon storage and reduced water production (Lindenmayer *et al.* 2011).

3. There are general principles for forest biodiversity conservation that are a valuable antidote to the problems associated with land use intensification

A major overarching theme in attempts to reduce the impacts of land use intensification on forest biodiversity must be to maintain habitat suitability for forest biota. This is because habitat loss is a key factor underpinning both the loss of biodiversity *per se* and the loss of species in forest ecosystems. Five general principles can help meet the critical goal of maintaining habitat across the full range of spatial scales. These are: (1) the maintenance of connectivity; (2) the maintenance of landscape heterogeneity; (3) the maintenance of stand structural complexity; (4) the maintenance of the integrity of aquatic ecosystems by sustaining hydrological and geomorphological processes; and (5) the use of knowledge of natural disturbance regimes in natural forests to guide off-reserve forest management practices (Lindenmayer and Franklin 2002).

Developing management plans and on-the-ground strategies to mitigate the effects of land use intensification and which are broadly consistent with the five general principles listed above (and hence consistent with the general overarching principle of habitat maintenance) requires three broad groups of conservation strategies: (1) setting aside large ecological reserves; (2) adopting landscape-level off-reserve management strategies; and (3) adopting stand-level off-reserve management strategies (Lindenmayer and Franklin 2002).

4. Land sparing in forests needs to be examined in terms of the quality of the land being spared, the kinds of species targeted for conservation and the conceptual landscape model that underpins decision making

Land sparing is where more intensive land use is adopted in a given location as a kind of ecological offset for increasing the area reserved for conservation in another location (Fischer *et al.* 2008; Green *et al.* 2005). Recent discussions about land sparing have been prominent in debates about agriculture and biodiversity conservation (Fischer *et al.* 2008). However, the idea actually has a longer history in the forestry sector than it does in agriculture (e.g. Gladstone and Ledig 1990). For example, Seymour and Hunter (1999) discussed the case of the maintenance of sustained yields of timber from the State of Maine in the USA. Maine has extensive (but relatively low intensity) wood production forestry and there are significant concerns about biodiversity conservation because few areas are in formal reserves. Seymour and Hunter (1999) noted that for every hectare of forest transferred into intensive forestry, between 3–5 hectares of forest could be set aside as ecological reserves and overall sustained yields of timber and pulpwood could be maintained.

Such an 'allocation solution' to production–conservation problems has been commonplace in forest environments for several decades (Lindenmayer and Franklin 2002). In many respects, allocation solutions would appear (at least superficially) to be a win for conservation and a win for wood production. There are no doubt circumstances where an allocation solution would indeed be a win-win outcome. However, careful analysis and appraisal are needed to determine whether the best outcome for conservation and production is a low intensity, widespread forest management approach or a reserve and intensively managed area approach. Such analysis demands an understanding of:

- The objectives of forest management practices and conservation efforts.
- The quality of the land that is 'spared'. This is critical as the land that is either currently in reserves or is proposed for protection often has fundamentally different environmental characteristics from those of land in production forests. Typically, reserved areas are steep, rocky or in areas of low productivity and which can be of significantly lower value for biodiversity than flatter, lower productivity wood production areas (Scott and Tear 2007). Thus, an increase in the size of reserved areas of such low productivity forest may not be an appropriate ecological offset for increased land use intensification in more productive areas.
- The kinds of species that are intended for conservation management. For example, if the aim of conservation efforts is to maintain species with limited mobility or that are dispersal or connectivity-limited, then the best option might be low intensity, widespread forest management rather than an allocation strategy based on reserves versus production areas (Franklin 1993).

Finally, the merit of the land sparing approach needs to be considered in the context of the landscape conceptual model which is being used to guide management actions and conservation efforts, including whether landscape cover is being considered from a human perspective or the perspective of particular organisms of interest (see Manning *et al.* 2004). The island model (MacArthur and Wilson 1967) of landscapes in which areas are deemed to be either habitat or non-habitat usually best corresponds to reserve/production area strategy. Conversely, low intensity, widespread forest management might be more congruent to those circumstances where landscapes are conceptualised using other kinds of models like the contour model or the continuum model in which different areas vary in habitat quality along gradients of suitability (Fischer *et al.* 2008). Notably, very few plans for managing forests and managing biodiversity consider the underlying landscape conceptual model but rather conceptualise landscapes solely from a human perspective and almost exclusively using the island model of vegetation cover (Lindenmayer *et al.* 2008).

5. Many of the problems associated with land use intensification are underpinned by deeper social and economic problems driven by the overcommitment of natural resources in landscapes

Many landscapes are overcommitted to various kinds of land uses such as livestock grazing, cropping, irrigation and urban development. There is insufficient 'ecological margin' to maintain ecologically sustainable environmental conditions or to stem further biodiversity loss. Water is overcommitted – flows for environmental purposes are often limited. The land is also overcommitted: areas required for production forestry, cropping or domestic livestock grazing mean that there is not enough land left for conservation. This has resulted in biodiversity loss, and has the potential to undermine the capacity of ecosystems to sustain wood production, maintain freshwater, and regulate climate and air quality.

Serious thinking, serious planning and serious on-ground management action are needed to avoid worsening the problems of already heavily overcommitted landscapes and forest resources. 'Landscape accounting' is needed to determine whether landscapes can realistically and ecologically support the intensification of industries and still maintain other key environmental values, maintain ecosystem services and support viable populations of native plants and animals. Ultimately, many kinds of landscape management problems cannot be tackled seriously until levels of resource allocation, and types and levels of resource use, are better planned. The problems of overcommitted landscapes can be magnified when events like major natural disturbances (e.g. bushfires, cyclones, hurricanes and droughts) significantly alter forest cover and the availability of the natural resources being utilised. As an example, many wood production areas in temperate regions around the world are managed on the basis of long-term sustained yields of timber and pulpwood. Yield calculations are generated from the area of forest available for harvesting but these rarely account for losses that will be incurred from natural disturbances like bushfires. When areas are lost to production but sustained yields remain unchanged (many regions have legislated levels of sustained yield), this can significantly increase harvesting pressure on undamaged areas with corresponding negative impacts on other values like the maintenance of key ecosystem services and biodiversity conservation. This is precisely what has occurred in the wet forests of south-eastern Australia following the 2009 bushfires in that region. The solutions to this kind of problem include: (1) the need to ensure that sustained yield calculations are sufficiently conservative to account for losses in resource availability that will invariably occur as a result of natural disturbances; and (2) the development of agility among policy makers and resource managers to allow them to respond rapidly to major natural disturbances and change both harvesting levels and the locations of where timber harvesting occurs.

6. There are always ecological surprises – the key is to learn how to better anticipate and detect them so they are not surprises

The lessons outlined above, together with the insights set out in the other chapters in this book, clearly indicate there is a lot of knowledge to inform the understanding of biodiversity responses to land use intensification. Despite this, as in all fields of work in ecology and natural resource management, studies of land use intensification are punctuated by many 'ecological surprises' – which can be broadly defined as unexpected findings about the natural environment (*sensu* Lindenmayer *et al.* 2010). Many of these unexpected findings are 'undesirable', including the unintended outcomes of management actions gone awry in intensively used landscapes. Improved understanding and better anticipatory capability for detecting ecological surprises is essential for creating human organisations and institutional arrangements that can better respond to unanticipated future events in a positive way. Lindenmayer *et al.* (2010) recently suggested a suite of approaches to bolster studies so that they might be better positioned to detect ecological surprises and generate the anticipatory capability needed to respond to them. These are highly relevant to landscapes subject to increasing land use intensification and they include (after Lindenmayer *et al.* 2010):

- Conducting a range of types of parallel and/or concurrent research in a given target area.
- Better using past literature and conceptual models of the target ecosystem in posing good scientific questions and developing hypotheses and alternative hypotheses.
- Increasing the capacity for ecological research to take advantage of opportunities arising from major natural disturbances.
- Maintaining existing, and instigating new, long-term studies.

Long-term studies, in particular, are important for detecting surprises as: (1) time is a major driver of change and some key phenomena do not become apparent unless targeted entities like populations, ecosystems and ecological processes are studied for a prolonged period; (2) increasing experience with a target ecosystem increases the chance that appropriate ecological questions can be posed and important new insights subsequently found, including ones that are unexpected; (3) experience with an ecosystem, ecological community or population heightens the potential to notice when something unusual or different occurs; and (4) longitudinal research will often allow the identification, quantification and deeper understanding of important phenomena that are not detectable with a cross-sectional approach.

Conclusions

Land use intensification is a major process threatening the conservation of biodiversity and the maintenance of key ecological processes in wood production forests worldwide and it can manifest as reduced stand structural complexity and increased landscape homogeneity. Strategies to address the potential problems of land use intensification include such traditional and well-known ones like the maintenance of stand structural complexity, connectivity and landscape heterogeneity. There are valuable cross-learning opportunities about the impacts of land use intensification from historical case examples and from empirical and theoretical work in forests and agricultural areas. Such cross-learning is critical to speed progress toward better integration of resource management and conservation management. Finally, there are ways to better design and implement empirical studies of land use intensification to improve the probability of detecting ecological surprises, especially those likely to have negative impacts on natural environments and the biota that inhabit them.

Acknowledgements

I thank my colleagues involved in co-organising the workshop from which this book was derived – in particular Andrew Young, Saul Cunningham and Linda Broadhurst. Claire Shepherd kindly assisted with many aspects of the organisation of that meeting and also assisted in the preparation of this chapter. Many of the ideas developed in this paper have arisen from long-term collaborative studies with Jerry Franklin, Gene Likens, Mac Hunter and Ross Cunningham.

Biography

David Lindenmayer is a Research Professor at The Australian National University. He has worked on biodiversity conservation and resource management issues for nearly three decades and published more than 760 scientific articles and over 30 books. He is a member of the Australian Academy of Science.

References

Angelstam P, Boutin S, Schmiegelow FA, Villard M, Drapeau P, Host G, Innes J, Isachenko G, Kuuluvainen T, Monkonnen M, Niemi G, Roberge J, Spence J and Stone D (2004) Targets for boreal forest biodiversity conservation – a rationale for macroecological research and internal co-operation. *Ecological Bulletin* **51**, 487–509.

Connell JH and Sousa WP (1983) On the evidence needed to judge ecological stability or persistence. *American Naturalist* **121**, 789–824.

Fischer J, Brosi B, Daily G, Ehrlich P, Goldman R, Goldstein J, Lindenmayer DB, Manning A, Mooney H, Pejchar L, Ranganathan J and Tallis H (2008) Should agricultural policies encourage land sparing or wildlife-friendly farming? *Frontiers in Ecology and the Environment* **6**, 380–385.

Franklin JF (1993) Preserving biodiversity – species, ecosystems, or landscapes? *Ecological Applications* **3**, 202–205.

Franklin JF and Forman RT (1987) Creating landscape patterns by forest cutting: ecological consequences and principles. *Landscape Ecology* **1**, 5–18.

Gladstone WT and Ledig FT (1990) Reducing the pressure on natural forests through high-yield forestry. *Forest Ecology and Management* **35**, 69–78.

Green RE, Connell SJ, Scharlemann JP and Balmford A (2005) Farming and the fate of wild nature. *Science* **307**, 550–555.

Gustafsson L, Kouki J and Sverdrup-Thygeson A (2010) Tree retention as a conservation measure in clearcut forests of northern Europe: a review of ecological consequences. *Scandinavian Journal of Forest Research* **25**, 295–308.

Lindenmayer DB (2009) Forest wildlife management and conservation. *Annals of the New York Academy of Sciences* **1162**, 284–310.

Lindenmayer DB and Franklin JF (2002). *Conserving Forest Biodiversity: A Comprehensive Multiscaled Approach*. Island Press, Washington.

Lindenmayer DB, Hobbs RJ, Likens GE, Krebs C and Banks S (2011). Newly discovered landscape traps produce regime shifts in wet forests. *Proceedings of the National Academy of Sciences* **108**, 15887–15891.

Lindenmayer DB, Hobbs RJ, Montague-Drake R, Alexandra J, Bennett A, Burgman M, Cale P, Calhoun A, Cramer V, Cullen P, Driscoll D, Fahrig L, Fischer J, Franklin J, Haila Y, Hunter M, Gibbons P, Lake S, Luck G, MacGregor C, McIntyre S, Mac Nally R, Manning A, Miller J, Mooney H, Noss R, Possingham H, Saunders D, Schmiegelow F, Scott M, Simberloff D, Sisk T, Tabor G, Walker B, Wiens J, Woinarski J and Zavaleta E (2008) A checklist for ecological management of landscapes for conservation. *Ecology Letters* **11**, 78–91.

Lindenmayer DB, Hunter ML, Burton PJ and Gibbons P (2009) Effects of logging on fire regimes in moist forests. *Conservation Letters* **2**, 271–277.

Lindenmayer DB, Likens GE, Krebs CJ and Hobbs RJ (2010) Improved probability of detection of ecological 'surprises'. *Proceedings of the National Academy of Sciences* **107**, 21957–21962.

Linder P and Östlund L (1998) Structural changes in three mid-boreal Swedish forest landscapes, 1885–1996. *Biological Conservation* **85**, 9–19.

MacArthur RH and MacArthur JW (1961) On bird species diversity. *Ecology* **42**, 594–598.

MacArthur RH and Wilson EO (1967) *The Theory of Island Biogeography*. Princeton University Press, Princeton.

Malhi Y, Aragao LE, Galbraith D, Huntingford C, Fisher R, Zelazowski P, Sitch S, McSweeney C and Meir P (2009) Exploring the likelihood and mechanism of a climate-change-induced dieback of the Amazon rainforest. *Proceedings of the National Academy of Sciences* **106**, 20610–20615.

Manning AD, Lindenmayer DB and Nix HA (2004) Continua and Umwelt: novel perspectives on viewing landscapes. *Oikos* **104**, 621–628.

Scott JM and Tear TH (2007) What are we conserving? Establishing multiscale conservation goals and objectives in the face of global threats. In *Managing and Designing Landscapes for Conservation*. (Eds DB Lindenmayer and RJ Hobbs) pp. 494–510. Blackwell Publishing, Oxford.

Seymour RS and Hunter ML (1999) Principles of ecological forestry. In *Maintaining Biodiversity in Forest Ecosystems*. (Ed. ML Hunter) pp. 22–61. Cambridge University Press, Cambridge.

14 INTENSIFICATION OF COFFEE PRODUCTION AND ITS BIODIVERSITY CONSEQUENCES

John Vandermeer and Ivette Perfecto

Lesson #1. Coffee agroforests can be major repositories of tropical biodiversity.

Lesson #2. The coffee agroecosystem contributes to a high-quality matrix.

Lesson #3. Intensification of coffee may lead to a breakdown of key spatial patterning with concomitant loss in ecosystem function.

Lesson #4. Biodiversity in the non-intensified system provides the key ecosystem service of pest control.

Lesson #5. Biodiversity enhances the ecosystem service of pollination.

Introduction

The traditional coffee agroecosystem has become iconic as an emblem of a managed ecosystem that can be planned so as to be sustainable in a variety of contexts. The best known of these contexts is biodiversity conservation. As the intensification of coffee production proceeds, first with the reduction of shade trees, then with accelerated use of agrochemicals and finally with the fully intensified sun coffee system, the consequences with respect to biodiversity can be severe (Perfecto *et al.* 1996). But those consequences are complicated, involving both the direct effects on biodiversity itself, but, perhaps more important, the indirect effects that biodiversity has on other aspects of the system, the so-called ecosystem services. First, there are the direct effects, the most celebrated of which is as a repository of biodiversity. The shaded coffee agroecosystem has been shown to have potentially enormous conservation value (Perfecto *et al.* 2007), sometimes housing biodiversity that rivals nearby natural systems (Perfecto *et al.* 1997). Additionally, the shaded coffee system can contribute a high quality matrix element for landscapes that maintain biodiversity through metapopulation or metacommunity dynamics (Vandermeer and Perfecto 2007; Perfecto and Vandermeer 2008a, 2010; Perfecto *et al.* 2009).

In addition to these direct effects of intensification on biodiversity, the biodiversity harboured within the coffee agroecosystem may indirectly contribute to some ecosystem services such as pollination (Jha and Vandermeer 2009, 2010) and pest control (Vandermeer *et al.* 2010). Intensification of the system may result in the reduction of the effectiveness of such ecosystem services.

In a completely different context, the coffee agroecosystem has been conceived of as a model for the study of more theoretical or 'pure' ecology (Vandermeer and Perfecto 2006; Greenberg *et al.* 2008; Perfecto and Vandermeer 2008a,b; Vandermeer *et al.* 2008; Philpott *et al.* 2008a,b,c). Although the details vary from place to place, the coffee agroecosystem represents one of the most common of terrestrial tropical ecosystems in the world and the study of basic ecological processes in it has led to conceptual advances that help to inform the basic theoretical formulations of biodiversity (Vandermeer *et al.* 2008, 2010a; Vandermeer and Perfecto 2006, 2007; Perfecto and Vandermeer 2008a). Indeed, it has been argued (Perfecto *et al.* 2008b) that the coffee agroecosystem be considered a 'model system' for community ecology. Because the same, or similar, practices can be found in all tropical areas of the world, it is an ecosystem that is effectively 'replicated' throughout the world, awaiting serious ecological study (Greenberg *et al.* 2008; Perfecto and Vandermeer 2008b). As such, studies of biodiversity within this 'model' system, have suggested new scientific questions about the underlying ecological theory associated with biodiversity (Vandermeer *et al.* 2008; Perfecto *et al.* 2008b; Vandermeer and Yitbarek 2012).

Lessons

1. Coffee agroforests can be major repositories of tropical biodiversity

We started our work on coffee on 1988 while teaching the Tropical Managed Ecosystem course for the Organization of Tropical Studies in Costa Rica. At that time the national government was stimulating the intensification of coffee farms in an attempt to increase productivity. The result was a vast agro-deforestation that eliminated many of the diverse coffee agroforests, replacing them with coffee monocultures. The ecological literature is overflowing with wondrous descriptions of the massive biodiversity in the tropics in general, and our own experience suggested to us that agroecosystems in the tropics could be similarly diverse, at least relative to their temperate zone cousins. Furthermore, we were intrigued by the way northern ecologists and conservation biologists made such a strong distinction between pristine ecosystems and human altered ecosystems, mainly agriculture. Ecological theory at the time suggested that intermediate levels of disturbances actually benefit biodiversity (Connell 1978; Huston 1979), a narrative that should have suggested that agricultural practices might somehow be related to biodiversity conservation. Nevertheless, in practice most conservation programs and even much research in pure ecology focused on natural and pristine ecosystems. The intensification of coffee plantations in Costa Rica provided us with the opportunity to study biodiversity within an agro-ecosystem and to ask the empirical question whether all agriculture is the same with respect to biodiversity, as seemed to be the tacit assumption among conservationists and many ecologists. Operationalising this question, we formulated the problem as the 'intensification gradient', and asked what the pattern of biodiversity loss was along this gradient, a focus that had a natural manifestation in the coffee system, even if it retains the problematic flavour of a non-operationalisable concept in other systems. Coffee is produced using

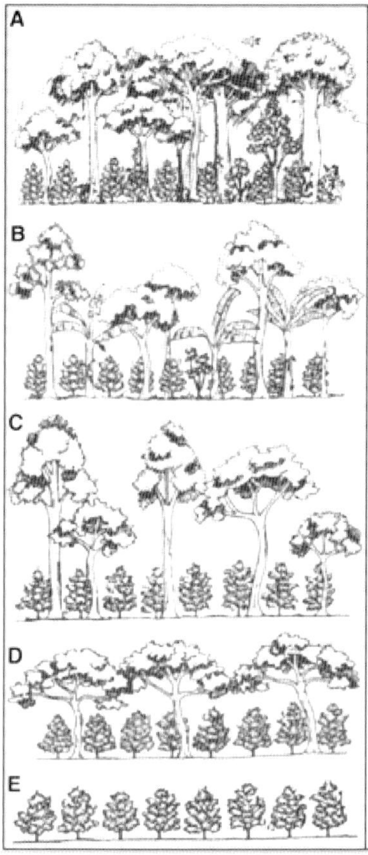

	MANAGEMENT SYSTEM	%SHADE* COVER	SHADE TREE* RICHNESS
A	RUSTIC	71-100	> 50
B	TRADITIONAL POLYCULTURE	41-70	21-50
C	COMMERCIAL POLYCULTURE	31-40	6-20
D	SHADED MONOCULTURE	10-30	1-5
E	UNSHADED (SUN) MONOCULTURE	0	0

Figure 14.1: Gradient of coffee management systems (Perfecto *et al.* 2005). *Figures for per cent shade and tree species richness are approximate based on field studies by ourselves and others.

a wide range of management systems going from the 'rustic' system where coffee is grown under a very diverse canopy of native forest trees (Figure 14.1A) to the sun system, which consists of coffee monocultures with no shade trees (Figure 14.1E). Our research was cast as one of biodiversity and agricultural intensification, using the coffee agroecosystem as a model system since it corresponds so well to the ideal of a gradient of intensification. The results of this research were somewhat surprising in that few researchers previously connected tropical biodiversity conservation in a positive way with agriculture. An early review paper (Perfecto *et al.* 1996) concluded that coffee agroforests could indeed be major sites for biodiversity conservation, and served as a springboard for the establishment of the Bird Friendly coffee certification program, stimulating much interest in the role of tropical agroforests in biodiversity conservation.

Since then, a vast literature has emerged on the role of agriculture in the conservation of biodiversity as well as on the mechanisms responsible for the maintenance of biodiversity within agroecosystems (Vandermeer and Perfecto 2007; Perfecto and Vandermeer 2008a). What has become clear is that there is no unitary pattern of biodiversity loss (Figure 14.2). Depending on the taxa involved, losses can be quite severe immediately after converting

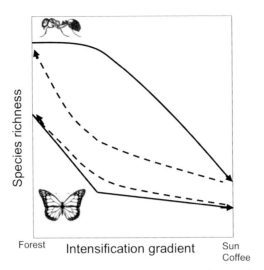

Figure 14.2: Observed pattern of species loss with coffee intensification for ants and butterflies (solid lines), and two hypothesised patterns of species recovery: one showing hysteresis (a different pathway of species recovery from species loss; dashed line for ants), and another with no hysteresis (same pathway for species loss and recovery; dashed line for butterflies). Data for species loss comes from Perfecto *et al.* (2003).

natural forest to coffee (see the butterfly data in Figure 14.2), but sometimes can be virtually unchanged until the system becomes excessively intensified (see the ant data in Figure 14.2). Indeed, it is now well recognised that the patterns of species loss with agricultural intensification can vary dramatically for different taxa and under different conditions (Philpott *et al.* 2008; Gordon *et al.* 2007).

2. The coffee agroecosystem contributes to a high-quality matrix

The landscape level processes that most ecologists now acknowledge operating in biodiversity dynamics require that both matrix and fragments be considered in any analysis of fragmented landscapes (which dominate the tropics today and probably will do so in the foreseeable future). Most organisms living in fragmented landscapes effectively live in the context of metapopulations, with the fragments housing each subpopulation of the metapopulation (Vandermeer and Perfecto 2007; Vandermeer *et al.* 2010a; Perfecto and Vandermeer 2008, 2010; Perfecto *et al.* 2009). At its most elementary level the equilibrium fraction of fragments occupied by the population is expected to be approximately $1-e/m$ (Levin's metapopulation model), where e is the extinction rate and m the migration rate. The value of m is frequently thought of as a function of the number and arrangement of habitat fragments in the landscape. It has been repeatedly emphasised in the literature (Perfecto and Vandermeer 2008, 2010) that the 'quality' of the matrix in which those fragments are located is likely to be a major determinant of the value of m. Consequently, the nature of the agroecosystems that comprise that matrix is important, not only as a potential repository of biodiversity, but also as dynamic patches in the landscape.

A variety of frameworks have been used in characterising landscapes in general (Turner *et al.* 2001; Forman 1995; Lindenmayer and Fischer 2006). A relatively new abstract framework includes what we refer to as propagating sinks, ephemeral sources and percolating mosaics (Vandermeer *et al.* 2010). This new framework is an attempt to incorporate a higher level of abstraction to metapopulation theory within the context of real managed landscapes. We refer to a habitat patch that allows for the temporary survival of a population, long enough to generate a new generation of propagules that move on to another patch, but not long enough to maintain the local population in perpetuity, as a 'propagating sink'. On the other hand, we refer to a habitat patch that regularly provides propagules to other habitats, but itself disappears eventually, as an 'ephemeral source'. We also refer to some combination of propagating sinks and ephemeral sources as a 'percolating mosaic'.

3. Intensification of coffee may lead to a breakdown of key spatial patterning with concomitant loss in ecosystem function

Since the report that the ant, *Azteca instabilis*, forms clusters of nests in shade trees that generate a spatial pattern suggestive of 'robust criticality' (Vandermeer *et al.* 2008), a great deal of attention has been given to the spatial dynamics of several components of the ecosystem connected to that species of ant, at several scales (Perfecto and Vandermeer 2008; Vandermeer *et al.* 2010; Philpott *et al.* 2008a,b; Perfecto and Vandermeer 2008; Jha and Vandermeer 2009). At a macro level, *A. instabilis* forms a distinctive pattern of patches of nests (e.g. in a 45-ha plot where we have been monitoring this species for six years), and those patches are basically formed by the interaction of the ant with a parasitic fly. The ants tend a scale insect (a potential coffee crop pest), which is prevented from reaching pest status in the farm as a whole by a lady beetle, which in turn is only able to persist because of the clusters of ant nests. Adding to this complexity is a fungal pathogen of the scale insect that only becomes epizootic when the scales are at a high density, which only happens in the large nest clusters. This entomopathic fungus also attacks the potentially devastating fungal disease – the coffee rust disease caused by *Hamelia vastatrix*. It has been argued that the management of four of the major potential pests of coffee (the coffee berry borer, the green coffee scale, the coffee rust disease and the coffee leaf miner) is actually accomplished autonomously through a network of ecological complexity. Through particular network structuring, some key nonlinearities, an element of stochasticity, and, especially, with the added dimension of space, our theoretical and empirical research (summarised in Vandermeer *et al.* 2010) reveals an ecological system persisting and generating ecosystem services as a result of complex interacting components, operative at a large spatial scale (many hectares).

Some of the other ants in the system seem to generate spatial pattern at the scale of 10s of metres through a totally distinct mechanism, simple interspecific competition. We have argued in the past that these pattern formations are examples of self-organisation, invoking such metaphors as the Turing process, or mosaic structures resulting from strong competition (Perfecto and Vandermeer 2006; 2008a). Regardless of the exact mechanism, it is clear that these other ants (1) exist in a spatial mosaic; and (2) are mainly predators. The fact that they are mainly predators means that they likely contribute to the control of potential pests. The fact that they exist in a spatial mosaic means that their community dynamics is likely an important component of their pest control impact, and this impact operates at a spatial scale that is considerably smaller than the *Azteca instabilis* scale mentioned above.

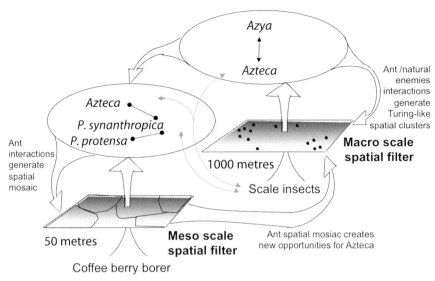

Figure 14.3: Proposed dynamic relationships between the two spatial scales and the predator/pest complex, where the beetle *Azya orbigera* is the predator on *Coccus viridis* (the scale insect) at the macro-scale, and the community of ants represents the predators on the coffee berry borer at the meso-scale. The 'predators' is a complex of species, either the Azya/Azteca/Phorid complex or the *Azteca/P. synanthropica/P. protensa* complex. In both cases there are many interactions involved among the various predatory species and their associates, some of which are indicated here with negative interactions symbolised by closed circles and positive interactions with arrowheads. The macro-scale filter refers mainly to our 45-hectare permanent plot and the small ovals represent the clumped Azteca nests as we have reported (Vandermeer *et al.* 2008). The meso-scale filter refers to areas of less than 1 hectare, containing the patchwork mosaic of different species of ants. *Pheidole synanthropica* is a species that nests in the ground but forages widely, both arboreally and terrestrially and represents the 'guild' of ants that forage similarly, most of which tend the scale insects. *Pheidole protensa* is a species that forages only on the ground and represents those species that forage similarly. Both interact with one another, with the many other ants in the system, and with Azteca.

These two spatial scales come together dialectically, as reflected in Figure 14.3.

4. Biodiversity in the non-intensified system provides the key ecosystem service of pest control

The classic notion of trophic structure is evident in the coffee system, with several herbivores eating coffee and those herbivores attacked by a variety of predators and parasites. Examining the trophic system composed of birds, spiders and insects in a large exclusion experiment, we conclude that the operation of the insurance hypothesis of biodiversity services is operative, and a special sort of trophic cascade is present. In particular, it appears to be the case that birds have a negative effect on spiders, more than on insects, including insects that may be potential herbivorous pests. The spiders involved are mainly orb weavers and are consequently major predators of small parasitoid wasps. The latter appear to be the main natural enemies of most herbivores in the system. Therefore, rather than the simple cascade from bird predator to herbivore, we appear to have a system where the birds reduce the spiders, thus releasing the parasitoids which provide most of the natural enemy control, bringing up the question of intra-guild predation and omnivory. More recent work (Williams-Guillen *et al.* 2008) has shown that here (as in all other bird exclusion studies), the experimental effect is actually exclusion of both birds and bats and that the latter have a stronger effect than the former in the wet season, when migrant birds are not present. Thus, the 'guild' – aerial vertebrate predators is the more appropriate appellation.

More recently we have also discovered trait-mediated cascading effects that result in an indirect facilitation of a coccinellid beetle by phorid parasitoids that attack the *Azteca* ants (part of the interacting network in Figure 14.3). The system is dramatically affected by the ability of various actors to detect the pheromones of the ants. The phorids locate the general vicinity of the ants by detecting the concentration of volatile pheromones they give off, but require movement by the ants in order to actually oviposit. As an apparent evolutionary response, the ant produces a special pheromone that signals the presence of a phorid, to which all ants in the vicinity respond by leaving the area or becoming catatonically still (since the phorid needs to see movement, this is an excellent strategy to avoid parasitisation). The coccinellid beetle has a larva that is protected by a waxy filamentous cuticle from attack by the ants and so is capable of eating large quantities of the ants' mutualist, the green coffee scale. However, there is an obvious problem faced by the coccinellid in that females are not able to approach the concentrations of scale insects that are needed by their offspring because of the protective activity of the swarming ants. In response, the beetle, when laden with eggs to be laid, has the ability to detect the special pheromone the ant produces when under attack by the phorids, obviously so as to take advantage of the fixed posture of the ants to oviposit in an area that is both protected from the ravages of the ants and among the high concentrations of their food source, the green scale insects. Therefore, there is an indirect effect (the forced reduction of ant activity by the phorid) that affects another indirect effect (the ant's protection of the scale insects against the beetle), which is to say a cascade of indirect effects.

For the ecosystem service of pest control this interaction web ultimately includes four coffee pests (the coffee berry borer, the green coffee scale insect, the coffee leaf miner and the coffee rust disease), that are controlled through indirect ecological interactions involving parasites, predators and pathogens, involving the ecological forces of competition, predation, parasitism, mutualism and pathogenicity, all in the context of a spatially extended system (Perfecto and Vandermeer 2006; Vandermeer *et al.* 2009, 2010; Larsen and Philpott 2010; Philpott *et al.* 2008). The relationship between pest control and matrix quality, although not as well documented as the relationship between matrix quality and biodiversity alone, nevertheless is suggested by these studies.

5. Biodiversity may enhance the ecosystem service of pollination

A substantial amount of work has been accomplished on the role of matrix quality for pollination services. Although the high quality *Coffea arabica* is self-pollinated, substantial evidence is now available indicating that pollinator activity increases seed set and ultimately yield (Klein 2009; Ricketts 2004). Part of the increase in yield is thought to be from the effects of cross-pollination, but an intriguing study involving ant exclusion (Philpott *et al.* 2006) demonstrates that with aggressive arboreal ants patrolling the branches, the positive effects of honey bees on pollination success is significant, presumably due to the ant's aggressive harassment causing the bees to move from flower to flower and bush to bush more frequently. It also may be the case that a diversity of native bees would provide a substantially more predictable beneficial pollination service, which would suggest that the high shade management of a system would be beneficial to this ecosystem service (Jha and Vandermeer 2009).

Conclusion

The coffee agroecosystem provides an excellent model system for the examination of biodiversity responses to changes in land use. It exists in an obvious range of intensification patterns (see Figure 14.1), in each of which there has been considerable research concerning biodiversity. Perhaps the most evident result of these investigations is that although biodiversity is lost through the process of intensification, the rate of that loss cannot be stated without stipulating the taxonomic categories of concern – birds respond differently from mammals, butterflies differently from ants. Beyond the biodiversity harboured in the agroecosystem itself, the contribution of coffee systems to the development of a high quality matrix in a fragmented landscape has been studied intensively with the conclusion that indeed the shaded coffee system does have an effect on creating a high quality matrix which leads to more biodiversity preservation at a landscape level, above and beyond the biodiversity contained within the agroecosystem itself.

We now understand that biodiversity indeed contributes to ecosystem function and thus the structure of the agroecosystem may have a major effect on the delivery of ecosystem services. For example, the bird biodiversity, known to be higher in shaded coffee systems compared to sun systems, contributes to the amelioration of insect outbreaks (Perfecto *et al.* 2004), therefore contributing to the ecosystem service of pest control. More subtly, the complex interaction of predators, prey, diseases and spatial structure can affect an 'autonomous' control over potential pests, illustrating how it is not just the number of species involved but also the diversity of the way in which different components of the biodiversity interact (Vandermeer *et al.* 2010).

Therefore, it is possible to conclude that shade management in coffee agroecosystems has a major impact on biodiversity in landscapes where coffee is grown, and that the consequent biodiversity may be intimately involved in the delivery of key ecosystem services. Transforming landscapes such that shade systems are reduced can therefore be expected to have a negative influence on biodiversity with concomitant loss of critical ecosystem services.

Biographies

John Vandermeer is the Asa Gray Distinguished University Professor of Ecology and Evolutionary Biology and the Arthur Thurneau Professor at the University of Michigan. He has been active in research on tropical forests, tropical agroecology and theoretical ecology for many years. He is author or editor of 13 books, most recently *The Ecology of Agroecosystems* (Jones and Bartlett, Sudbury Mass) and, with Ivette Perfecto and Angus Wright, *Nature's Matrix* (Earthscan, UK).

Ivette Perfecto is the George W. Pack Professor of Natural Resources at the University of Michigan. She has been active in research and analysis of agroecosystems and the politics of food and agriculture more generally. She is particularly focused on how local level multi-species interactions generate landscape level spatial patterns and the role such ecological forces play in the ecosystem service of pest control. Her most recent book is *Nature's Matrix* (Earthscan, UK), with co-authors John Vandermeer and Angus Wright.

References

Connell JH (1978) Diversity in tropical rain forests and coral reefs. *Science* **199**, 1302–1310.

Forman RT (1995) *Land Mosaics: The Ecology of Landscapes and Regions.* Cambridge University Press, NY.

Gordon C, Manson R, Sundberg J and Cruz-Angón A (2007) Biodiversity, profitability, and vegetation structure in a Mexican coffee agroecosystem. *Agriculture, Ecosystems & Environment* **118**, 256–266.

Greenberg R, Perfecto I and Philpott SM (2008) Agroforests as model systems for tropical ecology. *Ecology* **89**, 913–914.

Huston MA (1979) A general hypothesis of species diversity. *American Naturalist* **113**, 81–101.

Jha S and Vandermeer J (2009) Contrasting bee foraging in response to resource scale and local habitat management. *Oikos* **118**, 1174–1180.

Jha S and Vandermeer J (2010) Impacts of coffee agroforestry management on tropical bee communities. *Biological Conservation* **143**, 1423–1431.

Klein AM (2009) Nearby rainforest promotes coffee pollination by increasing spatio-temporal stability in bee species richness. *Forest Ecology and Management* **258**, 1838–1845.

Larsen A and Philpott S (2010) Twig-nesting ants: the hidden predators of the coffee berry borer in Chiapas, Mexico. *Biotropica* **42**, 342–347.

Lindenmayer DB and Fischer J (2006) *Habitat Fragmentation and Landscape Change.* Island Press, Washington.

López Gomez A, Williams-Linera G and Manson RH (2008) Vegetation structure and native and non-native tree species assemblages in coffee farms in an ecotonal region in Veracruz, Mexico. *Agriculture, Ecosystems & Environment* **124**, 160–172.

Perfecto I and Vandermeer J (2006) The effect of an ant/scale mutualism on the management of the coffee berry borer (*Hypothenemus hampei*) in southern Mexico. *Agriculture Ecosystems & Environment* **117**, 218–221.

Perfecto I and Vandermeer J (2008a) Biodiversity conservation in tropical agroecosystems: a new conservation paradigm. *Annals of the New York Academy of Science* **1134**, 173–200.

Perfecto I and Vandermeer J (2008b) Spatial pattern and ecological process in the coffee agroecosystem. *Ecology* **89**, 915–920 (Special Feature).

Perfecto I and Vandermeer J (2010) The agroecological matrix as alternative to the land-sparing/agriculture intensification model. *Proceedings of the National Academy of Sciences* **107**, 5786–5791.

Perfecto I, Vandermeer J and A Wright (2009) *Nature's Matrix: Linking Agriculture, Conservation and Food Sovereignty.* Earthscan, London.

Perfecto I, Armbrecht I, Philpott SM, Soto Pinto L and Dietsch TV (2007) Shade coffee and the stability of forest margins in Northern Latin America. In *The Stability of Tropical Rainforest Margins: Linking Ecological, Economic and Social Constraints*. (Eds T Tscharntke, M Zeller and C Leuschner) pp. 227–263. Springer-Verlag, Berlin.

Perfecto I, Mas A, Dietsche TV and Vandermeer J (2003) Species richness along an agricultural intensification gradient: a tri-taxa comparison in shade coffee in southern Mexico. *Biodiversity and Conservation* **12**, 1239–1252.

Perfecto I, Rice RR, Greenberg R and van der Voort ME (1996) Shade coffee: a disappearing refuge for biodiversity. *BioScience* **46**, 598–608.

Perfecto I, Vandermeer J, Hansen P and Cartín V (1997) Arthropod biodiversity loss and the transformation of a tropical agroecosystem. *Biodiversity and Conservation* **6**, 935–945.

Perfecto I, Vandermeer J, Mas A and Soto Pinto L (2005) Biodiversity, yield and shade coffee certification. *Ecological Economics* **54**, 435–446.

Perfecto I, Vandermeer JH, Lopez Bautista G, Ibarra Nunez G, Greenburg R, Bicher P and Langridge S (2004) Greater predation in shaded coffee farms: the role of resident neotropical birds. *Ecology* **85**, 2677–2681.

Philpott SM, Perfecto I and Vandermeer J (2008a) Effects of predatory ants on lower trophic levels across a gradient of coffee management complexity. *Journal of Animal Ecology* **77**, 505–511.

Philpott SM, Arendt W, Armbrecht I, Bichier P, Dietsch T, Gordon C, Greenberg R, Perfecto I, Soto-Pinto L, Tejeda-Cruz C, Williams G and Valenzuela J (2008b) Biodiversity loss in Latin American coffee landscapes: reviewing evidence on ants, birds, and trees. *Conservation Biology* **22**, 1093–1105.

Philpott SM, Uno S and Maldonado J (2006) The importance of ants and high-shade management to coffee pollination and fruit weight in Chiapas, Mexico. *Biodiversity and Conservation* **15**, 487–501.

Philpott SM, Lin BB, Jha S and Brines SJ (2008c) A multi-scale assessment of hurricane impacts on agricultural landscapes based on land use and topographic features. *Agriculture, Ecosystems & Environment* **128**, 12–20.

Philpott SM, Perfecto I and Vandermeer J (2008d) Behavioral diversity of predatory ants in coffee agroecosystems. *Environmental Entomology* **37**, 181–191.

Ricketts T (2004) Tropical forest fragments enhance pollinator activity in nearby coffee crops. *Conservation Biology* **18**, 1262–1271.

Turner M, Gardner RH and O'Neill RV (2001) *Landscape Ecology in Theory and Practice: Pattern and Process*. Springer, NY.

Vandermeer J and Perfecto I (2006) A keystone mutualism drives pattern in a power function. *Science*, **311**, 1000–1002.

Vandermeer J and Perfecto I (2007) The agricultural matrix and a future paradigm for conservation. *Conservation Biology* **21**, 274–277.

Vandermeer J, Perfecto I and Liere H (2009) Evidence for hyperparasitism of coffee rust (*Hemileia vastatrix*) by the entomogenous fungus, *Lecanicillium lecanii* through a complex ecological web. *Plant Pathology* **58**, 636–641.

Vandermeer J, Perfecto I and Schellhorn N (2010a) Propagating sinks, ephemeral sources and percolating mosaics: conservation in landscapes. *Landscape Ecology* **25**, 509–518.

Vandermeer J, Perfecto I and Philpott SM (2008) Clusters of ant colonies and robust criticality in a tropical agroecosystem. *Nature* **451**, 457–459.

Vandermeer J, Perfecto I and Philpott SM (2010b) Ecological complexity and pest control in organic coffee production: uncovering an autonomous ecosystem service. *BioScience* **60**, 527–537.

Vandermeer J and Yitbarek S (2012) Self-organized spatial pattern determines biodiversity in spatial competition. *Journal of Theoretical Biology* **300**, 48–56.

Williams-Guillen K, Perfecto I and Vandermeer J (2008) Bats limit insects in a neotropical agroforestry system. *Science* **320**, 70.

Part C

General discussion

15 PERSPECTIVES ON LAND USE INTENSIFICATION AND BIODIVERSITY CONSERVATION

David Lindenmayer, Saul Cunningham and Andrew Young

Introduction

There can be little doubt that there are truly colossal challenges associated with providing food, fibre and energy for an expanding world population without further accelerating already rapid rates of biodiversity loss and undermining the ecosystem processes on which we all depend. These challenges are further complicated by rapid changes in climate and its additional direct impacts on agriculture (see, for example, Lobell *et al.* 2011), ecosystem processes (Rockström *et al.* 2009) and biodiversity (Steffen *et al.* 2009). Nevertheless, it appears inevitable that there will be increased human demand for food, fibre and energy with corresponding pressure to transform many landscapes to new forms of land use (such as clearing native vegetation for agriculture) and to further intensify existing agricultural and forestry practices (Hodgson *et al.* 2010; Smith *et al.* 2010). For example, some major transitions in land use practices are already well under way such as the shift from grazing to cropping in parts of south-eastern Australia (Young and Broadhurst, Chapter 12). Expected local yield increases of an additional 2–3 tonnes of grain per ha could be realised by moving cropping into the high productivity–high rainfall zone in south-eastern and south-western Australia. In forestry, it is anticipated there will be a doubling of the plantation tree estate for the intensive production of industrial wood in the coming decades (Bauhus *et al.* 2010). In New Zealand, between 1960 and 2000, total agricultural production increased by 200–300% for major commodities with only a 6–7 % increase in land area farmed – that nation produces ~30% of the world's dairy products (Didham *et al.*, Chapter 9).

There are many different viewpoints about the best way to deal with the myriad issues associated with land use intensification – as demonstrated by the diversity of material in the different chapters in this book. Our aim in this final chapter is not to repeat the different perspectives of the chapter authors nor list the issues and problems they have raised. Rather, we focus on a subset of issues that have recommended themselves as being important by having appeared repeatedly throughout the chapters. We recognise that many of the topics that are examined in this chapter are interrelated, often intimately, but we have elected to treat them separately. In doing this, we accept some level of simplification has been necessary in the interests of clarity.

Definitions

The overarching theme of this book (and the meeting which gave rise to it) is land use intensification and its implications for biodiversity conservation. While some authors defined land use intensification (e.g. Didham *et al.*, Chapter 9; Lindenmayer, Chapter 13; Bawa *et al.*, Chapter 8) the majority did not and this reflects, at least to some degree, the difficulty in providing a definitive description that is useful across environments, production systems and land use objectives. In practice, the meaning of the term is context specific. For example, as noted in the contribution by Didham *et al.* (Chapter 9), the definition of land use intensification by Krebs *et al.* (1999) which has an ecological context, is markedly different from that of the Food and Agriculture Organisation (1995) which concerns production efficiency and has no ecological context. These differences can create considerable confusion, particularly when scale effects (see below) are ignored. For example, on-site increases in production efficiency may be driven by inputs derived off-site and for this reason there may be ecological impacts of land use intensification both on a given site as well as off-site (Didham *et al.*, Chapter 9). This is illustrated by the increase in dairy production in New Zealand that has been driven, in part, by the importation of products from a rapidly expanding Oil Palm industry in South-East Asia, but which in turn, has occurred through the clearing of rainforest (R. Didham, personal communication).

The multiple definitions of land use intensification also reflect different disciplinary perspectives, which can exaggerate conflict in the dialogue. If land use intensification is defined simply as increasing production from the same land, then from an agricultural point of view it would seem a necessary part of an agricultural solution. If land use intensification is defined as increasing transformation of the land away from the original habitat, then it would seem from a biodiversity perspective to epitomise the cause of the problem. In an attempt to reconcile these viewpoints, and improve the dialogue, some have promoted the concept of 'sustainable intensification' (Benton, Chapter 4), which supposes that intensification of production does not always have to come at a high biodiversity cost – the solution is to find these better approaches that increase production, but are sustainable in terms of impacts on biodiversity.

It is important to appreciate the implications of these different definitions of land use intensification and how they shape attitudes, but here we deliberately side step the problem of choosing a single favoured definition. Rather, we acknowledge that all kinds of land use change affects biodiversity to different degrees, and focus on understanding the kinds of change that have historically posed the greatest threats to biodiversity, and then how future modes of production might reduce these negative impacts. Such a concept necessarily includes the broad spectrum of changes from localised intensification through increasing yields within current land use boundaries to increases in productivity through conversion of new land to agriculture, and many combinations in between.

Scale

Topics associated with spatial scale arose repeatedly throughout almost all of the chapters in this book. For instance, multi-scaled impacts of land use intensification and attempts to mitigate those impacts were a key theme in discussions about land sparing versus land sharing spectrum of choices (see below). Many other, sometimes quite disparate, issues associated with scale were raised by different chapter authors.

McIntyre (Chapter 6) and Didham *et al.* (Chapter 9) described how land use intensification at a site level can create larger-scale problems (see also Hodgson *et al.* 2010) through, for example, movements of nutrients that lead to declines in native plants. Thus, increasing levels of fertiliser might have negative impacts on measures of on-farm biodiversity like plant species richness (e.g. Kleijn *et al.* 2009), but there also may be larger-scale effects on species richness through landscape-level degradation in vegetation condition and cover (McIntyre, Chapter 6). Didham *et al.* (Chapter 9) argued there was often limited information about off-site impacts created by on-site intensification (e.g. intensification of farming impacts on adjacent nature reserves) and this should be a key area of future research, particularly because this will create new insights into the mechanisms driving ecological changes in production landscapes. Benton (Chapter 4) argued that the problem of off-site impacts highlighted the need for landscape-scale management. In general, it was acknowledged that the spatial scale at which management actions are appropriate can vary depending on the organisms and management activities in question (see Wiens *et al.* 1997). Unfortunately, the best scale for managing good biodiversity outcomes is often greater than the scale upon which property rights are partitioned, and implies a need for mechanisms that reward coordination across property boundaries and even up to regional scales. This was clearly demonstrated by several of the chapters that reported data on the spatial dynamics of key ecological processes important for biodiversity conservation (Young and Broadhurst, Chapter 12 and Villard, Chapter 10).

Another scale issue that was touched on in a number of chapters was that of cross-scale influences. That is, actions which attempt to carry out ecologically sustainable practices in a given area but which may create problems elsewhere (Didham *et al.*, Chapter 9; Villard, Chapter 10). For example, increased conservation efforts in a given area can lead to intercontinental shifts in sources of timber provided to the market (see below) – this represents classic displacement of demand. In other cases, land use transformation can be underpinned by large-scale (often global) economic drivers. Thus, local actions can be driven by global market mechanisms but these can have landscape, region, national or even international effects (Didham *et al.*, Chapter 9; Young and Broadhurst, Chapter 12). These kinds of cross-scale phenomena can be very hard to develop management responses to because the scale of drivers, the scale of management and the scale of effects are all different. Market tools (e.g. certification schemes) and regulatory instruments can help direct the market towards better biodiversity outcomes and have worked in some contexts (kiwifruit – Didham *et al.*, Chapter 9; timber in South Africa – Samways, Chapter 7). In landscapes that are geared to production for export (such as Australia and New Zealand), there is a case for the development of new ways to facilitate planning and conservation beyond the farm scale. There are also regions, however, where local actions are much less connected with global drivers (because of distant markets and insufficient capital). This occurs in the

Darjeeling region of India and this disconnect can have positive outcomes for ecologically sustainable land management, including the conservation of biodiversity, as it has the potential to place the responsibility and authority to manage both production and biodiversity into the same hands. Nevertheless, even in these cases, there appears to be a strong case for matching the scales of management and impact to the scales of governance and land use planning (Bawa *et al.*, Chapter 8).

In summary, scale issues pervade almost all topics associated with land use intensification including its impacts on other values (like biodiversity conservation), and approaches to mitigate those effects. An understanding of scale is critical for quantifying on-site and off-site effects of intensification, mitigating those effects where they manifest, and understanding the interacting influence of mis-matches in the scale of drivers of change, the scale of management and the resulting scale of effects. Scale is particularly prominent in debates about the efficacy (or otherwise) of land sparing versus land sharing approaches to commodity production and biodiversity conservation – the primary topic of the following section.

Land sparing

At the smallest scales, the focus of conservation decision making is restricted to what happens within a given plot of land, be it under productive use or otherwise. But at larger geographic scales, one can conceptualise a choice between extensive but low input land uses, or mosaics with contrasting patches of intensive production and conservation lands. These options, which can only be conceived at landscape scale or above, can be described as the 'land sharing versus land sparing' spectrum, with the extremes characterised by the 'sparing' model, where more intensive land use (leading for example to increased crop yields) is adopted in a given location as a kind of ecological offset for increasing the area reserved for conservation in another location (Fischer *et al.* 2008: Green *et al.* 2005: Phalan *et al.* 2011), and 'sharing' where production and biodiversity conservation goals are integrated in the same location. Many authors (including Benton, McIntyre and others) have reflected on these choices in their chapters.

Interestingly, there has developed an increasing sense of contentious debate in the literature about the sharing versus sparing options (Ewers *et al.* 2009; Fischer *et al.* 2008; Green *et al.* 2005; Phalan *et al.* 2011). While the debate has been at times oppositional, the starting point, that land use decisions and their consequences are best understood at larger scales, is widely accepted. The debate has focused on the sense there are advocates for one or the other end of the spectrum, and that one or the other strategy is inherently better. To characterise some of the main concerns, critics of land sparing argue that while the offset principle is possible in theory, in practice it usually does not occur – that is intensification occurs, but no conservation land is released (Ewers *et al.* 2009). Further, the emphasis on intensification risks eroding the key ecosystem processes which underpin production (Garibaldi *et al.* 2011) and thereby undermines attempts to boost yields (Tscharntke *et al.*, Chapter 2). Critics of land sharing argue that the mixed strategy compromises both good conservation, and good production, giving you a landscape that is less effective on both fronts.

While the debate in the workshop echoed some of this contention, it was agreed that the best outcome for any given landscape was in fact shaped by the particularities of the system.

In other words, we know of landscapes where land sharing might lead to good production and conservation outcomes, such as some tropical agroforests (Tscharntke, Vandermeer and Perfecto). Equally, we know of landscapes where the contrast between the dominant production practice and the original native vegetation is so great, that land sparing would seem more likely to produce a better outcome. Importantly, it also was recognised that there is no need to treat the choices as dichotomous – a land use plan for good conservation and production outcomes may well use elements of both ends of the spectrum, such as McIntyre's conception of a mosaic blending vegetation remnants, grazed native vegetation and cropping (McIntyre, Chapter 6). Critically, this means that making good land use choices needs good information on the relationship between production systems and biodiversity outcomes in the focal landscape, rather than adopting a one-size-fits all policy.

Research to guide better management

The various chapters in this book make it clear there are many issues and problems associated with the impacts of land use intensification on biodiversity, as well as strategies for mitigating those impacts (such as land sparing; see discussion above). Developing better management strategies for the future will require a deeper understanding of the consequences of land use change for biodiversity and production within a given system. The production of coffee is possibly the best-developed model ecosystem currently available (Vandermeer and Perfecto, Chapter 14). This is because a range of different production systems are employed across a gradient of production intensities – from low-intensity shade coffee through to high-intensity full-sun coffee. Vandermeer and Perfecto (Chapter 14) have demonstrated through a number of impressive studies that there is a clear response to a land use intensification gradient in coffee production systems. However, the effects of land use intensification vary substantially between species and among sets of taxa (Vandermeer and Perfecto, Chapter 14). Interestingly, the work by Vandermeer and Perfecto suggests that for some tropical ecosystems like those where coffee production is a key form of land use, the land sharing approach may well be an important and viable alternative to land sparing (see the preceding section). This is a similar conclusion to that reached by Tscharntke (Chapter 2) based on a major body of work in tropical cacao production systems where, like coffee production, the intensity of management regimes can vary markedly in space (and time).

Few other ecosystems have the elegant gradient of production intensities that characterise coffee and cacao production. Yet, research in other ecosystems can provide useful models for generating important insights about land use intensification. For example, investigations in Canadian forests, summarised in Chapter 10 by Villard, illustrate a number of key biotic responses to the amount and spatial configuration of intensively logged forest, partially logged forest and unlogged areas. Responses ranged from altered connectivity and elevated levels of nest predation in birds, to altered abundance and fecundity of lichens.

Another model ecosystem is the heavily altered temperate woodland ecosystems in south-eastern Australia. Studies in these systems by McIntyre (Chapter 6) have underscored how changes in nutrient levels can lead to major ecosystem degradation. Young and Broadhurst (Chapter 12) and Cunningham *et al.* (Chapter 11) emphasised the importance of multi-disciplinary studies in these temperate woodland ecosystems for understanding the impacts of land used intensification on biodiversity and key processes. The body of work summarised in both of those chapters emphasised the need to integrate demographic and genetic research as well as the critical importance of social and economic insights in driving land use change and landscape conservation and management.

In many respects, almost all of the chapters in this book are based on detailed empirical information drawn from studies of particular ecosystems. These particular studies then create the traditional dilemma in ecology and resource management of knowing how to extrapolate lessons from a specific set of circumstances to more general situations (Lindenmayer *et al.* 2008). That is, developing a robust framework for making inference about likely biological effects in a range of places and contexts. Model systems are useful in that they often are underpinned by a deep understanding of ecosystem responses drawn from many different studies conducted in the same location. However, few systems can be subjected to a range of detailed studies and the key challenge is to know how to use information from such well-understood places to guiding appropriate decision making in less well-researched places.

Critical importance of connectivity

Dispersal-limited species are among those most at risk of decline or extinction as a consequence of land use change (reviewed by Lindenmayer and Fischer 2006). Given this, a key part of conserving biodiversity in production landscapes is the maintenance of connectivity (Lindenmayer and Franklin 2002; Lindenmayer, Chapter 13). The importance of connectivity was emphasised in many chapters in this book (e.g. Vandermeer and Perfecto, Chapter 14; Villard, Chapter 10; Young and Broadhurst, Chapter 12). For example, Samways (Chapter 7) highlighted the value of what he termed 'ecological networks' for the integration of biodiversity conservation and commodity production from several studies in southern Africa. Similarly, the maintenance of connectivity was pivotal to the persistence of biota in tropical production landscapes in Latin America, with the permeability of the 'matrix' for movement deemed to be crucial (Vandermeer and Perfecto, Chapter 14). Young and Broadhurst (Chapter 12) identified thresholds in connectivity for plants in temperate agricultural landscapes in south-eastern Australia and noted that despite very high levels of landscape modification there were sometimes still high levels of functional connectivity. Based on ecological theory, McIntyre (Chapter 6) suggested that levels of connectivity in a landscape could be directly linked with levels of cover of native vegetation; connectivity begins to be most impaired once 30% of the original cover of native vegetation is removed (McIntyre, Chapter 6).

The maintenance of appropriate connectivity is an important component of any effective conservation strategy in production landscapes. However, the task of maintaining or enhancing connectivity is not necessarily a straightforward one because the concept of connectivity is complex and multi-faceted. In an attempt to tackle this complexity, Lindenmayer and Fischer (2007) recognised three kinds of connectivity. *Habitat connectivity* can be broadly defined as an emergent property of ecosystem mosaics that reflects the influence of landscape structure on a species' mobility and its probability of survival within and among resource patches. The concept of *landscape connectivity* is based on a human perspective of landscape pattern and typically relates to the connectedness of native vegetation patches. In some circumstances, and for some species, habitat connectivity and landscape connectivity will be closely correlated. In others, habitat connectivity for a given species will be quite different from a human perspective of landscape connectivity. *Ecological connectivity* is the connectedness of key ecosystem processes (Lindenmayer and Fischer 2007). Understanding these relationships is crucial because we generally manage on the basis of habitat connectivity, with the goal of achieving landscape connectivity.

Attempts to maintain connectivity to reduce the impacts of land use intensification need to be prefaced by questions like: *What kind of connectivity? Connectivity for what?* And, *Connectivity at what spatial scale?* Answers to these questions are important because connectivity is different for different organisms and for different ecological processes and can sometimes have undesirable consequences (Lindenmayer and Fischer 2007). This, in turn, influences whether efforts to maintain or increase connectivity are best addressed through establishing corridors, stepping stones or a continuous cover of trees (or other vegetation) (see Franklin 1993) or some other kind of approach like large, contiguous, regional-scale areas (Worboys *et al.* 2010). The options for creating connectivity in production landscapes are strongly influenced by the *contrast* between the production and non-production patches in the landscape and the dispersal abilities of various elements of the biota. Where the contrast is high (e.g. forest versus crop) and mean mobility is low, connectivity for most organisms will be restricted to corridors or stepping stones. Where the contrast is low (e.g. agroforesty system versus forest; grazed woodland versus natural woodland) and mobility is high, there will be much greater scope for managing the whole landscape for permeability, with a less dichotomous view of the different landscape patches (Vandermeer and Perfecto, Chapter 14; McIntyre, Chapter 6).

Predictive ability for species responses to land use intensification

The ability to accurately predict those species at risk of decline or extinction as a result of land use intensification is an important part of informed and proactive conservation actions in production environments. McIntyre (Chapter 6) suggested that the responses of some species to land use intensification were predictable based on an understanding of biogeographic history. For example, she argued that bird and plant taxa naturally associated with wetlands would be robust to some kinds of intensive land uses like irrigated rice cropping because of similarities in characteristics of disturbance regimes. Notably, creating congruence between human disturbance regimes and natural disturbance regimes is a long applied general principle in the conservation management of natural forests (Hunter 1993: Korpilahti and Kuuluvainen 2002: Swanson *et al.* 2011). The underlying ecological thinking is that organisms are likely to be best adapted to the disturbance regimes under which they have evolved, but are potentially susceptible to novel forms of disturbance (or combinations of disturbances) such as those that are more or less frequent and/or more or less intensive than would normally occur (Hunter 2007). Other aspects of biogeographic history are critically important in their influence on predicting species responses to land use intensification. Attwood and Burns (Chapter 3) highlighted the need to understand the time scale on which land use transitions have occurred (e.g. New Zealand and Australia versus the United Kingdom) and how this context has a big effect on what fauna and flora have persisted and what management options are available for integrating conservation with production.

Studies of extinction proneness provide another avenue of work that shows promise for predicting the potential impacts of land use intensification and, in turn, informing proactive management strategies. Such investigations aim to identify the attributes of sets of species that are extinction prone. For example, Sodhi et al. (Chapter 5) discussed the kinds of species at high risk of being lost through rainforest clearing in South-East Asia. The advantage of advance knowledge of the extinction proneness of species is that it can help suggest pre-emptive actions to prevent problems developing and which are more likely to be successful because they can be enacted before a 'crisis' has developed. The problem with studies of extinction proneness is that they show that different species with markedly different life history attributes are vulnerable to decline in the same landscape and in different landscapes (reviewed by Lindenmayer and Fischer 2006). This may limit predictive ability in extinction proneness among different groups in the same landscape and the predictive ability for the same groups between different landscapes (Attwood and Burns, Chapter 3). This has been the experience from long-term work in the intensive wood production plantation landscapes in southern New South Wales (Lindenmayer 2009). Similarly, work by Vandermeer and Perfecto (Chapter 14) on 'model systems' like those used for coffee production show clear responses to an intensification gradient. However, how such effects manifest varies markedly between groups of taxa (Vandermeer and Perfecto, Chapter 14) with winners and losers of all kinds.

The concept of extinction debt takes the idea of extinction proneness, and then highlights that the final moment of loss for these species is often delayed long after the critical land use change has occurred. For example, an extinction debt of between 5 and 21% has been paid on National Parks in North America after they were gazetted (Carroll et al. 2004). While some intensive production landscapes like those in the agricultural areas of southeastern Australia still support high levels of biodiversity (Young and Broadhurst, Chapter 12; Cunningham et al. Chapter 11), it remains unclear how many of those species are at risk of future extinction. The converse of the extinction debt concept is that of invasion credit. That is, a lot of invasive species in agricultural areas may be yet to invade existing areas of remnant native vegetation because these species have only recently become established. Didham et al. (Chapter 9) argued that the relatively recent colonisation of potentially invasive plants in native vegetation in New Zealand is likely to be in 'invasion credit'. This contrasts with the remnant native vegetation of Europe where the system has been exposed to invasive plants for a very long time (Didham et al., Chapter 9). The implications of invasion credits are ultimately the decline and loss of native taxa as invasive species become established and spread in the remaining fragments of native vegetation.

Another issue brought forward by several authors is the importance of understanding the ecological processes behind observed species responses to agricultural intensification. These processes are what explain the differential vulnerability that is often observed, but can predict likely levels of extinction debt. A lack of process dynamics knowledge, which has already been shown to be important in terms of scale issues, will result in perverse management outcomes when apparently ecologically similar species respond differently to the same management actions due to modest differences in reproductive ecology, or dispersal ability.

Despite the challenges in developing accurate predictions of extinction-proneness in response to land use intensification, it is important to continue this work. Additional approaches, such as those based on general landscape ecology principles also may provide useful guidance on the likely responses of biota to land use intensification. For example, patch size is a well-known and valuable predictor of plant and animal response. This is because large patches support large populations with high levels of fecundity and recruitment (Young and Broadhurst, Chapter 12). This suggests a key part of attempts to integrate conservation and production in multi-use landscapes will be maintaining the integrity of large patches of remnant native vegetation (Young and Broadhurst, Chapter 12). However, a focus on large patches does not mean that small patches are without conservation value – these often can be important, especially when they are in good condition and have high levels of functionality (Fischer and Lindenmayer 2002; Da Silva *et al.* 2011).

Conclusions

It is clear that there are many important and often complex dimensions to decision making associated with land use intensification and biodiversity conservation. The reality is that we have been able to only briefly touch on a subset of issues in this book. There was insufficient space (and expertise) to examine some of the significant drivers of land use intensification like rates of population growth, changing patterns of food consumption (e.g. massive increases in the amount of diary food eaten in Asia: Robert 2008), the availability of fossil fuels or the impact of biofuel production (for example Fargione *et al.* 2010). We therefore have not explored some potential solutions including human population control, changes in the ways food is distributed, changes in patterns of food consumption (such as reducing the amount of meat eaten and increasing the number of insects and vegetables consumed; Westing 2010) or reducing levels of food wastage. Rather, we have focused attention on whether science can suggest better approaches to reducing the conflict between productive land use and biodiversity conservation. The chapters in this book indicate there are some important lessons from the work conducted on land use intensification to date.

First, in many production landscapes, much of the land is privately owned, regardless of whether or not it is currently being used for agriculture (e.g. New Zealand, see Didham *et al.*, Chapter 9) and hence the application of traditional conservation strategies like setting aside large ecological reserves are typically not possible. Off-reserve schemes are therefore required, with the agri-environment schemes of Europe and North America (e.g. Perkins *et al.* 2011) and the recently implemented Environmental Stewardship Program in Australia being examples (Attwood and Burns, Chapter 3). These programs can sometimes deliver positive conservation outcomes in production landscapes (e.g. Perkins *et al.* 2011) and a key part of achieving their aims will be for conservation biologists to help determine how to create a better alignment of production and conservation interests (Benton, Chapter 4). A second key part of the success of these programs will be for scientists to better communicate the results of their research to policy makers and resource managers to facilitate the adoption of new information into on-the-ground management (Attwood and Burns, Chapter 3; Young and Broadhurst, Chapter 12; see Lindenmayer *et al.* 2011). Third, despite the good intentions of agri-environment schemes, their effectiveness remains poorly known (e.g. European Court of Auditors 2005 in Whitfield 2006; UK Parliament 2010; Attwood and Burns, Chapter 3), although recent work suggests they can generate some positive outcomes for biodiversity (e.g. Perkins *et al.* 2011). A key problem with agri-environment schemes to date has been a lack of well-developed criteria for measuring success, coupled with a paucity of well-designed and implemented ecological monitoring programs to quantify progress and return on investment (Carwell 2010). Clearly then, tackling problems with land use intensification demands the development of a tractable suite of indicators of management success (not a single bioindicator taxon; Samways, Chapter 7) that is accompanied by rigorous monitoring programs.

A second lesson from work to date on land use intensification is the need to avoid irreversibility. That, is, circumstances in which landscapes are so radically altered that it is almost impossible to restore them (Cunningham *et al.*, Chapter 11) because they have been permanently shifted into an alternative stable state (*sensu* Connell and Sousa 1983). The problems of irreversibility are associated not only with the modification of landscapes and severely impaired ecosystem processes (Cunningham *et al.*, Chapter 11; McIntyre, Chapter 6; Tscharntke *et al.*, Chapter 2) but also future extinction debts, and future exotic species credits (see Didham *et al.*, Chapter 9).

Several chapter authors argued that new institutions will be required to meet the challenges posed in feeding a greater human population in an ecologically sustainable way without undermining the integrity of key ecosystem processes or further accelerating species loss (see also Reganold *et al.* 2011). These new institutions will be needed to: (1) help coordinate better and more directed scientific endeavours associated with solving complex sustainability issues in production landscapes; (2) better support the kinds of monitoring programs needed to guide ecologically sustainable land use practices (Bawa *et al.*, Chapter 8; Bawa 2010; Lindenmayer and Likens 2011); and (3) identify ways to link resource management scales with the scale of initiatives aimed at delivering effective impact mitigation (Young and Broadhurst, Chapter 12).

Finally, it is clear that access to food, the maintenance of key ecosystem processes and the conservation of biodiversity are global problems that are creating major land management challenges for humans living in almost all areas of the planet (Sodhi *et al.*, Chapter 5). However, given marked differences in the scale and type of agricultural systems, the natural history and ecology of agricultural regions, land tenure, markets and biotic assemblages, we argue that local, context-specific solutions will be most appropriate for solving this global problem. For example, approaches to balancing agricultural production and biodiversity conservation in the Darjeeling region of India (Sodhi *et al.*, Chapter 5) will be markedly different from those attempts to tackle the same broad kind of problems in meso-American tropical systems (Vandermeer and Perfecto, Chapter 14) and different again from the industrial-scale agricultural production areas of south-eastern Australia (Cunningham *et al.*, Chapter 11; McIntyre, Chapter 6; Young and Broadhurst, Chapter 12) and mid-western North America. Such context-specific solutions underscore the critical importance of good science (and the good communication of that science) to guide policy development, decision-making and ultimately on-the-ground management. Our hope is that the contributions in this book can assist people to make good integrated environmental, social and economic decisions about how land is used and how to avoid or mitigate the negative environmental impacts of future land use changes.

Acknowledgements

We thank Claire Shepherd for her help with many aspects of the preparation of this chapter. Linda Broadhurst and a number of colleagues provided useful comments that improved earlier versions of this chapter.

References

Bauhus J, van der Meer P and Kanninen M (2010) *Ecosystem Goods and Services from Plantation Forests.* CSIRO Publishing, Melbourne.

Bawa K (2010) Monitoring systems outdated and protectionist. *Nature* **446**, 920.

Carroll C, Noss RF, Paquet PC and Schumaker NH (2004) Extinction debt of Protected Areas in developing landscapes. *Conservation Biology* **18**, 1110–1120.

Carwell M (2010) Rural development in the United Kingdom: continuity and change. *International Journal of Land, Law and Agricultural Science* **4**.

Connell JH and Sousa WP (1983) On the evidence needed to judge ecological stability or persistence. *American Naturalist* **121**, 789–824.

Da Silva F, Do Prado V and Rossa-Feres D (2011) Value of small forest fragments to amphibians. *Science* **332**, 1033.

Ewers R, Scharlemann JP, Balmford A and Green RE (2009) Do increases in agricultural yield spare land for nature? *Global Change Biology* **15**, 1716–1726.

FAO (2004) Planning for sustainable use of land resources. *FAO Land and Water Bulletin* **2**.

Fargione JE, Plevin RJ and Hill JD (2010) The ecological impact of biofuels. *Annual Review of Ecology, Evolution, and Systematics* **41**, 379–406.

Fischer J, Brosi B, Daily G, Ehrlich P, Goldman R, Goldstein J, Lindenmayer DB, Manning A, Mooney H, Pejchar L, Ranganathan J and Tallis H (2008) Should agricultural policies encourage land sparing or wildlife-friendly farming? *Frontiers in Ecology and the Environment* **6**, 380–385.

Fischer J and Lindenmayer DB (2002). Small patches can be valuable for biodiversity conservation: two case studies on birds in southeastern Australia. *Biological Conservation* **106**, 129–136.

Franklin JF (1993). Preserving biodiversity – species, ecosystems, or landscapes? *Ecological Applications* **3**, 202–205.

Garibaldi LA, Aizen MA, Klein AM, Cunningham SA and Harder LD (2011) Global growth and stability of agricultural yield decrease with pollinator dependence. *Proceedings of the National Academy of Sciences* **108**, 5909–5914.

Green RE, Connell SJ, Scharlemann JP and Balmford A (2005) Farming and the fate of wild nature. *Science* **307**, 550–555.

Hodgson JA, Kunin WE, Thomas CD, Benton TG and Gabriel D (2010) Comparing organic farming and land sparing: optomizing yield and butterfly populations at a landscape scale. *Ecology Letters* **13**, 1358–1367.

Hunter ML (1993) Natural fire regimes as spatial models for managing boreal forests. *Biological Conservation* **65**, 115–120.

Hunter ML (2007) Core principles for using natural disturbance regimes to inform landscape management. In *Managing and Designing Landscapes for Conservation: Moving from Perspectives to Principles*. (Eds DB Lindenmayer and RJ Hobbs) pp. 408–422. Blackwell Publishing, Oxford.

Kleijn D, Kohler F, Báldi A, Batáry P, Concepción ED, Clough Y, Díaz M, Gabriel D, Holzschuh A, Knop E, Kovács A, Marshall EJP, Tscharntke T and Verhulst J (2009) On the relationship between farmland biodiversity and land-use intensity in Europe. *Proceedings of the Royal Society B* **276**, 903–909.

Korpilahti E and Kuuluvainen T (2002) Disturbance dynamics in boreal forests: defining the ecological basis of restoration and management of biodiversity. *Silva Fennica* **36**, 1–447.

Krebs JR, Wilson JD, Bradbury RB and Siriwardena GM (1999) The second silent spring? *Nature* **400**, 611–612.

Lindenmayer DB (2009) *Large-Scale Landscape Experiments. Lessons from Tumut*. Cambridge University Press, Cambridge.

Lindenmayer DB, Archer S, Barton P, Bond S, Crane M, Gibbons P, Kay G, MacGregor C, Manning A, Michael D, Montague-Drake R, Munro N, Muntz R, Okada S and Stagoll K (2011) *What Makes a Good Farm for Wildlife?* CSIRO Publishing. Melbourne.

Lindenmayer DB and Fischer J (2006) *Habitat Fragmentation and Landscape Change*. Island Press, Washington, D.C.

Lindenmayer DB and Fischer J (2007) Tackling the habitat fragmentation panchreston. *Trends in Ecology and Evolution* **22**, 127–132.

Lindenmayer DB and Franklin JF (2002) *Conserving Forest Biodiversity: A Comprehensive Multiscaled Approach*. Island Press, Washington.

Lindenmayer DB and Likens GE (2011) Effective monitoring of agriculture. *Journal of Environmental Monitoring* **13**, 1559–1563.

Lindenmayer DB, Hobbs RJ, Montague-Drake R, Alexandra J, Bennett A, Burgman M, Cale P, Calhoun A, Cramer V, Cullen P, Driscoll D, Fahrig L, Fischer J, Franklin J, Haila Y, Hunter M, Gibbons P, Lake S, Luck G, MacGregor C, McIntyre S, Mac Nally, R, Manning A, Miller J, Mooney H, Noss R, Possingham H, Saunders D, Schmiegelow F, Scott M, Simberloff D, Sisk T, Tabor G, Walker B, Wiens J, Woinarski J and Zavaleta E (2008) A checklist for ecological management of landscapes for conservation. *Ecology Letters* **11**, 78–91.

Lobell DB, Schlenker W and Costa-Roberts J (2011) Climate trends and global crop production since 1980. *Science* **333**(6042), 616–620.

Perkins AJ, Maggs HE, Watson A and Wilson JD (2011) Adaptive management and targeting of agri-environment schemes does benfit biodiversity: a case study of the corn bunting *Emberiza calandra*. *Journal of Applied Ecology* **48**, 514–522.

Phalan B, Balmford A, Green RE and Scharlemann JPW (2011) Minimising harm to biodiversity of producing more food globally. *Food Policy* **36**, S62–S71.

Reganold JP, Jackson-Smith D, Batie SS, Harwood RR, Kornegay JL, Bucks D, Flora CB, Hanson JC, Jury WA, Meyer D, Schumacher A, Sehmsdorf H, Shennan C, Thrupp LA and Willis P (2011) Transforming U.S. agriculture. *Science* **332**, 670–671.

Robert P (2008) *The End of Food*. Bloomsbury, New York.

Rockström J, Steffen W, Noone K, Persson Å, Chapin IFS, Lambin E, Lenton TM, Scheffer M, Folke C, Schellnhuber H, Nykvist B, De Wit CA, Hughes T, van der Leeuw S, Rodhe H, Sörlin S, Snyder PK, Costanza R, Svedin U, Falkenmark M, Karlberg L, Corell RW, Fabry VJ, Hansen J, Walker B, Liverman D, Richardson K, Crutzen P and Foley J (2009) Planetary boundaries: exploring the safe operating space for humanity. *Ecology and Society* **14**, 32.

Smith P, Gregory PJ, van Vuuren D, Obersteiner M, Havlik P, Rounsevell M, Woods J, Stehfest E and Bellarby J (2010) Competition for land. *Philosophical Transactions of the Royal Society B Biological Sciences* **365**, 2941–2957.

Steffen W, Burbidge A, Hughes L, Kitching R, Lindenmayer DB, Musgrave W, Stafford-Smith M and Werner P (2009) *Australia's Biodiversity and Climate Change*. CSIRO Publishing, Melbourne.

Swanson ME, Franklin JF, Beschta RL, Crisafulli CM, DellaSala DA, Hutto RL, Lindenmayer DB and Swanson FJ (2011) The forgotten stage of forest succession: early-successional ecosystems on forest sites. *Frontiers in Ecology and the Environment* **9**, 117–125.

United Kingdom Parliament (2010) HC 611 The outcome of the Comprehensive Spending Review – supplementary written evidence submitted by the Department for Environment, Food and Rural Affairs. (CSR 01A).

Westing AH (2010) Food security: population controls. *Science* **328**, 169.

Wiens JA, Schooley RL and Weekes RD (1997) Patchy landscapes and animal movements: do beetles percolate? *Oikos* **78**, 257–264.

Whitfield J (2006) How green was my subsidy? *Nature* **439**, 908–909.

Worboys GL, Francis WL and Lockwood M (Eds) (2010) *Connectivity Conservation Management: A Global Guide*. Earthscan, London.

Index

Advances in Agroecology

Series Editor: Clive A. Edwards

Agroecosystems in a Changing Climate, Paul C.D. Newton, R. Andrew Carran, Grant R. Edwards, and Pascal A. Niklaus

Agroecosystem Sustainability: Developing Practical Strategies, Stephen R. Gliessman

Agroforestry in Sustainable Agricultural Systems, Louise E. Buck, James P. Lassoie, and Erick C.M. Fernandes

Biodiversity in Agroecosystems, Wanda Williams Collins and Calvin O. Qualset

The Conversion to Sustainable Agriculture: Principles, Processes, and Practices, Stephen R. Gliessman and Martha Rosemeyer

Global Economic and Environmental Aspects of Biofuels, David Pimentel

Integrated Assessment of Health and Sustainability of Agroecosystems, Thomas Gitau, Margaret W. Gitau, and David Waltner-Toews

Interactions between Agroecosystems and Rural Communities, Cornelia Flora

Land Use Intensification: Effects on Agriculture, Biodiversity, and Ecological Processes, David Lindenmayer, Saul Cunningham, and Andrew Young

Landscape Ecology in Agroecosystems Management, Lech Ryszkowski

Microbial Ecology in Sustainable Agroecosystems, Tanya Cheeke, David C. Coleman, and Diana H. Wall

Multi-Scale Integrated Analysis of Agroecosystems, Mario Giampietro

Soil Ecology in Sustainable Agricultural Systems, Lijbert Brussaard and Ronald Ferrera-Cerrato

Soil Organic Matter in Sustainable Agriculture, Fred Magdoff and Ray R. Weil

Soil Tillage in Agroecosystems, Adel El Titi

Structure and Function in Agroecosystem Design and Management, Masae Shiyomi and Hiroshi Koizumi

Sustainable Agriculture and New Biotechnologies, Noureddine Benkeblia

Sustainable Agroecosystem Management: Integrating Ecology, Economics and Society, Patrick J. Bohlen and Gar House

Tropical Agroecosystems, John H. Vandermeer